# A New Way of Being

# A New Way of Being

Encounters, Engagements,
Explorations, and Applications

edited by Jack Petranker

Dharma Publishing

PERSPECTIVES IN TIME SPACE AND KNOWLEDGE

Dimensions of Thought I and II
Mastery of Mind
Visions of Knowledge
Light of Knowledge
A New Way of Being
A New Kind of Knowledge

Library of Congress Control Number 2004100294

ISBN 0-89800-356-3



Cover design by Emily Jan. Labyrinth design based on the labyrinth at Chartes Cathedral.

Typeset in Adobe Trump Mediaeval and Helvetica Light. Printed and bound by Dharma Press, Berkeley, CA.

9 8 7 6 5 4 3 2 1

Dedicated to Being

and to being differently

# Contents

## ENGAGEMENTS

## EXPLORATIONS



## APPLICATIONS

# Preface

*Beneath the surface of separate appearances, everything in the universe connects to space, heart to heart.*
—Tarthang Tulku,
Sacred Dimensions of Time and Space

Does a vision have a history? In one sense, the answer is certainly yes. The TSK vision first found written expression twenty-five years ago, with the publication of *Time, Space, and Knowledge: A New Vision of Reality*. Since then, four more books have been published: *Love of Knowledge* (1987), *Knowledge of Time and Space* (1990), *Dynamics of Time and Space*, and *Sacred Dimensions of Time and Space*. Another five volumes have been published in the Perspectives on TSK series, starting with *Dimensions of Thought* (in two volumes, 1980) and continuing with *Mastery of Mind* (1993), *Visions of Knowledge*, (1993), *Light of Knowledge* (1997), and now two additional volumes: *A New Way of Being* and *A New Kind of Knowledge*.

Nor is that all. The first TSK seminar, held several months before the original book was published, has been followed by a steady stream of programs that present and explore TSK, including workshops, classes, retreats, and long-term intensive residential programs. Then there is the history of how individuals have engaged the vision, a history known best by the individuals who have formed it.

There is a different way to tell the story. History in its ordinary sense depends on a conventional understanding of time as linear progression—exactly the model for understanding time that the TSK vision calls into question. If vision connects to the dynamic of time directly, without the 'this then that' structure we are used to tracing out, does it make sense to speak of history at all?

There is no need to choose between these two modes, nor is there any need to answer the question posed at the outset. A book like this one, which marks an anniversary in linear time, certainly has history on its side. But a book true to TSK will want to point beyond history as well. Whether this book meets that test will be up to the reader to decide.

The tendency to see two sides to every question is characteristic of TSK. Yet TSK does not 'take' sides. It sets them aside, looking for knowledge through a different form of inquiry.

A good example of this dynamic, one that bears directly on what the TSK vision is all about, can be found in *Mind over Matter*, published in 2002. There

Tarthang Tulku reflects on the potential applicability of TSK to our times. He identifies two sides in the modern way of thinking, but chooses neither:

> As a secular way of thinking increasingly calls into queston the belief systems that once governed this society, there have been two major responses. One response (prevalent in certain religious groups) is to withdraw from contact with those outside the circle of believers. But there are also those thinkers and practitioners who attempt to bring their own beliefs into harmony with the views and beliefs of the society they inhabit. On the one hand, the dividing lines or edges between competing sets of beliefs become more sharply defined; on the other hand, more and more people move into the great open spaces between these narrow enclaves, attempting to shape beliefs and values that could be considered universal.
>
> TSK could be seen as a teaching for these open spaces. It insists on no moral principles, founds no temples, initiates no priests, and relies on nothing remotely resembling prayer or worship.

At this point, it seems that Tarthang Tulku has chosen sides after all. TSK is for the open spaces. It takes no stands save one: a stand for what is open, free from all limits. But the same passage goes on to undermine this interpretation:

> [T]he openness that shapes TSK operates in a different dimension. TSK rejects the dichotomy

> between 'open' and 'closed' and the separation of 'inside' from 'outside'. From a TSK perspective, beliefs do not emerge to shape reality from some inaccessible, privileged place. Knowledge is already available, free to present itself by taking any form whatsoever.
>
> For those who might otherwise be wary of any claims to 'higher' knowledge, this approach effectively dissipates such concerns, suggesting that we could replace knowledge 'of', in all its competing flavors, with a more global knowledge 'in'. And it locates the source of this knowledge not somewhere else, but in a playful questioning of the knowledge upon which we already rely. In practical terms, this means that individuals are not bound to any hierarchy. They can adopt with equal facility the role of teacher or student, observer or participant.

The special strength of this volume (and its companion volume, *A New Kind of Knowledge*) is that it embodies the last sentence of the passage just quoted. The twenty-three contributors to this volume take on the roles of teacher, student, observer, and participant, all at once. They show what it means not to be "bound to any hierarchy," and in this way they pay tribute to the TSK vision in action.

Not bound to any one role, the writers represented here go beyond themselves. They make clear that the way of knowing TSK celebrates and activates is also a way of being, based not on identity or substance, but on space,openness, and (paradoxically?) not-knowing.

One way to conceive of a TSK way of being is in terms of time, space, and knowledge themeselves. At the close of *Love of Knowledge* (1987), Tarthang Tulku offers three successive exercises that give substance to this thought. Instead of putting the concerns of the self at the center of experience, he suggests, try putting space at the center of experience. What does this mean in practice? Is there any sense that doing this could truly give rise to a new way of being? Now do the same with time, and then with knowledge. Even some casual experiments in this direction can be very suggestive.

For readers interested in this approach, the theme of 'conducting' time, developed in *Dynamics of Time and Space*, may prove especially fruitful. The conductor conducts the orchestra, but the conductor also conducts the music, in the sense of letting it flow through his or her own heart and mind and out into the world. Are these two ways of being, two different ways of engaging time and space and knowledge?

This question, like all TSK questions, is best explored through practice. TSK comes alive for those who practice it, and they come alive as well. Through their practice they explore, they challenge, they question, and they apply. None of this is remote; none of it is confined to the realm of intellectual excitement that fuels the world of academic publishing. Whether practice happens in exercises, in daily life, or in actively engaging ideas, it aims at how we are in the world, at 'what's happening' and 'the way things are.'

To practice TSK is to go beyond, which is what the writers whose articles, thoughts, reflections, and dialogs are collected here do. Going beyond, they share what they find: not results but the lay of the land, the feel of the field, the flow of this special conducting.

Because it comes alive in this away, the present book (like its companion volume) has the potential to serve as a good introduction to the TSK vision and TSK spirit. You will find here few summaries or reconfigured presentations of 'established' TSK themes. The authors engage TSK, and if you engage what they have written, you will be well on the way to practicing TSK. Explore and practice as you read; loosen up your adventure muscles and enjoy the ride. Then you will be true to the TSK vision. You will be participating in the most fitting way in this celebration of what TSK has given the world till now and may yet give it in the future.

# Abbreviations

The following works in the *Time, Space, and Knowledge* series or the *Perspectives on Time, Space, and Knowledge* series, listed in order of publication date, have been abbreviated as indicated. All are published by Dharma Publishing. Note that the abbreviation "TSK" is sometimes also used in this volume to refer to the Time, Space, and Knowledge vision. It should be clear from context which usage is intended.

Works by Tarthang Tulku

| | |
|---|---|
| TSK | *Time, Space, and Knowledge: A New Vision of Reality*. 1977. |
| LOK | *Love of Knowledge*. 1985. |
| KTS | *Knowledge of Time and Space*. 1990. |
| VOK | *Visions of Knowledge: Liberation of the Modern Mind*. 1993. |

| | |
|---|---|
| DTS | *Dynamics of Time and Space: Transcending Limits on Knowledge.* 1994. |
| SDTS | *Sacred Dimensions of Time and Space.* 1997 |

Other Works

| | |
|---|---|
| DOT | *Dimensions of Thought: Explorations in Time, Space, and Knowledge* (2 volumes), eds. Ralph Moon and Steve Randall. 1980. |
| MOM | *Mastery of Mind*, ed. Jack Petranker, 1993. |
| LOK | *Light of Knowledge*, ed. Jack Petranker. 1997. |

# ENCOUNTERS

The relation between knowledge and being is not readily understood in our times. We are used to thinking of knowledge as a possession, something one has access to, like a sandwich we stuff into our backpack before we set out on a hike.

That picture, however, is far too limiting. In the history of civilizations, new knowledge has been a source of transformation, both at the social and personal levels. That is why this volume begins with accounts of personal encounters with the TSK vision. The power of that vision lies not only in its vast theoretical sweep, though there is much there to celebrate and explore, but also in its power to affect those who come in contact with it, changing their way of understanding reality, and ultimately their way of being in the world.

The first article in this section is by the late Bill Soskin, a long-time student of Tarthang Tulku (whom he refers to as "Rinpoche," a title of respect for Tibetan teachers). It appeared first in the Fall 1977 issue of *Gesar* magazine, shortly before the publication of *Time, Space, and Knowledge*. Inspired by a TSK seminar given by Tarthang Tulku at Odiyan, a retreat center just then opening its doors for the first time, it communicates beautifully the sense of excitement that the

TSK vision generated in its first audience. Bill modestly writes that he could not follow the discussions as well as other participants more thoroughly trained in philosophy, but it is clear from his presentation that the experience at the seminar changed him in profound ways and helped him deepen his own path of inquiry.

Toward the close of his article, Bill Soskin wonders whether the TSK vision will develop as an entirely new path, or whether it will find a place within the traditions of Tibetan Buddhism. In the course of TSK's subsequent development, Tarthang Tulku addressed this issue repeatedly, always emphasizing that the TSK teachings should not be seen as a restatement of Buddhism. (For an especially compelling discussion of what TSK is not, see DTS, pp. xvi–xx.) Yet in reading this early account of a TSK seminar and its impact, we are reminded that knowledge, though it expresses its power through many forms, remains one.

The next article, by Ronald Purser, brings the story of TSK's unfolding up to date. The author's involvement with TSK began more than twenty years ago. More than most students of the TSK vision, he has actively worked to integrate TSK into his professional and academic interests, and in this article he introduces the reader to some of these developments. He ends the article by returning to the theme that also closed Bill Soskin's reflections: the link between TSK and the Buddhist heritage of Tarthang Tulku.

Piet Hut's article identifies three ways of knowing: bottom-up, level, and top-down. Students of TSK might not all agree on where TSK fits within this scheme, or whether it fits at all. What make Piet's presentation so engaging is his way of setting this theoretical structure in play alongside and together with his own personal adventure of discovery. In the Postscript to the article, Piet describes the circumstances of his first encounter with TSK, his deepening connection, and his sense of how his inquiry will unfold in the future.

The next several articles, some quite short, link TSK to the sense of wonder. As Helena Antonova notes, we all experience such wonder as children. For her, the link is very concrete, for she identifies specific TSK exercises with games she played in childhood. Meindert Gijzen starts out from the same link between TSK and childhood experience, but broadens his inquiry to map out the special qualities of the TSK vision (or "virus"). Along the way he makes a useful distinction between TSK and the countless "Inner Technologies of Freedom" that are so widespread in these dissatisfied and hungry times.

Ruth Fellhauer takes up and condenses the theme of wonder, focusing on those moments of discovery and exploration that can emerge at any time. Inviting us to engage such moments, she writes simply, "Let's dance!" Ken McKeon, in an article orginally written as a review of *Dynamics of Time and Space,* expands what the previous article condensed, evoking in particular the ways that DTS engages language and invites an exploration of light.

Just as Ken McKeon looks primarily to *Dynamics of Time and Space,* so Lee Nichol directs his inquiry toward *Sacred Dimensions of Time and Space.* His reflections might be considered somewhat more abstract than those that have gone before, except that his article challenges the distinction between abstract and concrete, theoretical and practical. He offers "here I am" as the central theme of *Sacred Dimensions* and suggests that the special force of TSK has to do with engaging what that book calls "the prior." The reader may want to compare the distinction that Meindert Gijzen makes between TSK and what he calls "Inner Technologies of Freedom."

The final article in this section, by Charles T. Davis, III, chronicles a series of personal encounters that led the author to deeper sources of knowledge, including TSK. Several of the other articles have explored the relation between TSK and Buddhism, but here Davis writes without hesitation or equivocation of the connection he finds between TSK and his Christian faith.

# "Well, Now, What Did You Think?"

William F. Soskin

Very late in the Saturday night session at Odiyan, after he had spent many intense hours earlier that day and the preceding evening laying out for us the general framework of his time, space, and knowledge "theory," Tarthang Tulku turned to me as he had to one or two others immediately before and asked, "Well, now, what did you think?"

I knew the feeling. He had been pouring out to us for a day and a half the distilled product of several years of thought and now he wanted feedback. Almost at once I realized I had been preparing for that question. If Rinpoche had not asked now for some statement that would provoke further discussion, there was certainly already an insistent prior "audience" within me, demanding of myself some statement about what I had learned, what I thought, what my reactions were.

In answer to his question, I could only tell Rinpoche and my fellow seminar members that after a day and a half of intent listening I was in a state of excited bewilderment and creative unknowing. I surely understood far too little of contemporary astrophysical theory to know how to grapple with his ideas from that standpoint, while on the other hand his journeys into "inner space" were so different from anything I had ever imagined that I felt not unlike a fifteenth-century dockside roustabout in Genoa listening to the tales of mariners newly returned from across the sea. Rinpoche's ideas, shared over the course of the seminar, were so wholly new to me, so daring, so far beyond my immediate capacity to assimilate and analyze and evaluate that I needed time to ponder, to catch up at my own more plodding pace. But I did have other first reactions that seemed inappropriate to share just then. So I will share them now.

Once during the seminar, in a stretch of free time as I walked alone during the weekend looking out over the ocean that spreads below the Odiyan rim structure, I recalled a day like this from the distant past, a time shortly after I had completed my Ph.D. and was restlessly feeling my long and too narrow preoccupation with psychology. A chance browsing in the campus bookstore at the University of Chicago, where I was teaching at the time, led me to pick up Sir Arthur Eddington's then new book, *The Nature of the Physical World*. I took it home and within hours found myself drawn to a whole new world stage. In the weeks and months that followed, like one malnourished, I turned avidly from Eddington to other semi-popular works by Schrödinger and Hans Reichenbach and Asimov and Gamow, on to Heisenberg, and finally to a fascinating play with the "information theory" that Shannon had generalized out of the earlier work of Enrico Fermi, then himself a familiar sometimes habitué of the stand-up hotdog lunch counter at the back of the University bookstore. . . .

I remembered now the origin of this train of thought. Somewhere in that reading were vivid descriptions of the consternation that greeted Einstein's great disclosures. The world of physics split wide. A whole bloc of noted professors scoffed and denounced, while others pondered privately with varying degrees of doubt. When I was a boy it was still "good press" for reporters to write front page stories about the brilliant European Jewish physicist whose theories were so profound that supposedly only a handful of scientists in the entire world could understand them. Yet, within a generation or so, "young Turks" barely into the shaving stage were playing with ideas of relativity as though they had learned them in their nursery rhymes.

Here at Odiyan was a similar situation. Ideas like these that Rinpoche was expressing startle and dazzle and shock at first. It takes a while to feel at home with them. Rinpoche talked of small time and Great Time, of small space and Great Space, and afterwards I recalled a splendid starry night in a distant warm July. I was standing beside a giant reflecting telescope on the roof of the Yerkes Observatory in southern Wisconsin, on whose grounds we happened to be conducting a social experiment.

One midnight the observatory director came to my little laboratory room and invited me to the rooftop. In the darkness we felt our way up ladders and along narrow catwalks and finally arrived at a huge canvas-covered machine for piercing the night sky, a great reflecting telescope. Quickly, deftly, the elderly Dutch astronomer aimed and focused his precious machine, then stepped backward, and with a terse phrase and flourish, invited me, "Now, look!"

I had been gazing in quiet admiration a moment before at the million scintillating pinholes in the timeless canopy overhead, and now in the darkness I groped downward for the eyepiece, to inspect his wares. Suddenly with breathless amazement I found myself staring into the very heart of Andromeda! Beyond even the Aurignacian bunters, beyond Earth, beyond sun and moon and stars, almost beyond time itself. The very spectacle I was staring at no longer existed, a mere apparition whose image had sped across a million years of space and time to fall upon this mirror this splendid summer evening.

I trembled. My mind whirled. Beyond its overwhelming beauty, I had no way to absorb and assimilate what I was looking at. I stared and stared, and afterward I wanted to be alone. My companion understood. Quietly he left me to sit in solitude on the rooftop for a long stretch of that warm July night, staring inward, outward, upward, nowhere, everywhere, in stunned amazement.

Here on the hard floor of the new Odiyan, with often aching legs folded under me, I was being tempted back into that space and time by Rinpoche's talks. The admixture of his colorfully labored sentences and the force with which he stretched language to fit concepts of galactic scope and heroic boldness produced a gossamer world of images and schemata that flared and faded, pulsed in rhythms of understanding and obscurity.

Another time at Odiyan I remembered with amusement Mark Twain's delightful yarn, "Captain Stormfield's Visit to Heaven." The Old Salt fell from a yardarm one sunny afternoon as his ship lay at anchor in San Francisco Bay, and in an instant of painless transformation his soul was on its way out over the Golden Gate. After some daring high jinks with asteroids in the far reaches of outer space, the Captain finally reached the towering walls of Heaven, there to find himself before impediments so immense and gates so high that as he gazed upward they seemed almost to disappear into the clouds. Once or twice in the Saturday afternoon session at Odiyan, at moments when my mind could no longer grasp the subtleties of Rinpoche's distinctions between small space and Great Space, I felt as though I were beside Captain Stormfield outside some cosmic gate, whose great hinges would creak open for an ever-so-fleeting moment, and—as if through the eyepiece of the Yerkes telescope—I would glimpse a breathtaking world beyond. My wits would sharpen, my pulse quicken, and then the giant hinges would swing shut again, and I was no longer sure what I had seen, or thought I understood.

What did Rinpoche say? What had I understood? I was impressed with the scholarly-sounding questions and reactions propounded by some of my colleagues at the seminar who were obviously more steeped than I in physics and philosophy, whose familiarity with the concepts and discourse of epistemology kept me racing to follow the quick exchanges. I knew that it would surely take me weeks and months, if not years to follow Rinpoche even a few steps within those gates, beyond those high walls, to recapture the excitement of new-born ideas but also to linger in the quiet knowingness of the fleeting intimations that ebbed and flowed.

The very next morning, after returning home, I sat in my living room poring over some tables of data from a computer printout. My mind drifted back to the seminar, to play again with the concepts of a space filled with limitless forms, a "reality" that was no more than an arbitrary fixing of selected forms, a time that admitted of no past or future, but whose every discrete moment contained all pasts and futures, movement into either direction of which melted into a myriad of nows, themselves devoid of direction or continuity. A Knowledge that exists independent of any . . . .

As I struggled to focus my thoughts a fleeting recollection came of a thanka depicting the Bodhisattva Avalokiteshvara, with ten heads and a thousand arms, "each endowed with the eye of knowing." Small wonder, I thought, that Rinpoche, raised among such images, gamboled so lithely across space and time.

"Well, now, what did you think?"

My excited bewilderment has subsided gently since the seminar. It has also deepened, and has spawned moments when new inner vistas seem to bloom and fade in turn. But there is the feeling that a new perspective may be taking shape. Ever since meeting Rinpoche five or six years ago, I have been led periodically into a succession of wholly unanticipated adventures of consciousness, each one going far beyond its predecessor, each one a stunning new insight. The word "adventure" is entirely appropriate, and not the least trivializing. There has crept into my awareness a quality of recurrent awe that envelops me at irregular intervals, that taps me to an awakening realization that my lived time is only a fraction of the time I live, that my "behavior" and "work" and daily being are only one frame, one dimension of my total existence.

If anyone pressed me to say more explicitly what I mean, I might try to grope for phrases and concepts, but it would be more out of politeness. All the while I would prefer that I had not been asked, would prefer to remain quiet and say nothing, because the words that I might speak seem to have very little direct connection with my experience. Even the scientist in me now lives comfortably with such inchoate thoughts and ineffable experiences.

When I trace carefully from my very limited knowledge of the development of Buddhist thought through Hinayana and Mahayana to the exasperatingly little I understand of Vajrayana, I sense that the three are closely related, a rather natural progression of exercises and disciplines and ideas, all stepping stones along Man's unending path toward greater understanding. In Tibetan Buddhism if one allows oneself to be too concrete or literal one enters a seeming morass, a concrescence of fragments of all the paths of Man's past exploration into consciousness. The arch seems to span all the way from magic and the most elementary forms of "religious" expressions to what are perhaps the most daring and sophisticated explorations ever ventured. Very, very slowly it has dawned on me that all these tools and exercises and disciplines and forms represent a full array of resources, both museum and workshop of the tools of inquiry with which Man over millennia has sought to probe his own Being.

I feel myself ever so slightly wiser now, wise enough to recognize that brash young physicists playing with their black holes and quarks may be involved in a Hegelian dialectic or time warp, or an excursion along a Möbius strip circling out across the millennia, gradually drifting backward past medieval alchemists to wander once more along paths of consciousness in which ancient Pythagoreans sensed their place in a Grand Design. And in the process I realize that somehow I have been helped to allow myself to tarry a while amidst the incense and icons and mantras, long enough to discover with respectful appreciation that all these tools are essential to discovery and yield their full measure of reward, if only we do not too insistently ask ourselves such foolish, pseudo-questions as, "Is there a God?" "Did Buddha ever live?"

"Well, now, what do you think?"

I think that out of the generations and centuries of contemplation and disputation that the predecessors and teachers of Vajrayana devoted to their very special science of consciousness in India and Tibet there developed a remarkable body of understanding, unparalleled in the main currents of contemporary Western thought. I think that the few Tibetan high lamas I have encountered or whose works I have read convey clearly the sense that as a group those who are initiated into the higher teachings have direct access to and control of phenomena not known or appreciated or acknowledged in Western society or Western science. I think that collectively and individually the surviving Tibetan refugees are venturing to expose long-secret knowledge and capacities to Western scholars and students.

I think, too, that there may be an important difference between Eastern and Western systematic inquiry. The former has steadfastly explored space and time and knowledge in close relation to Being, and seems to move inexorably toward an understanding that ennobles the human condition as a special phenomenon in nature. Whereas Western science—still in wary reaction to the limitations and excesses of medieval Western religious philosophy—often struggles to develop its scientific equations and systems within a framework that arbitrarily blocks out the essence of the human equation.

I think this, Rinpoche: Your Time, Space and Knowledge seminar helped me to glimpse a mite more deeply into the contemplations of the lineages of Vajrayana, and to follow along gropingly toward your own next step in this long development. The weekend was an admirable and daring exercise, like the boldness of your still-forming thoughts, the very kind of audacious extensions that a Buddha would have delighted in. Thus it struck me that your topic was the most fitting possible for a first seminar at Odiyan. Seven or eight years ago, when you first arrived in Berkeley, you dreamed of a country place, a new asylum for an exiled body of refugee scholars, a new rooting place for a homeless religion. But years later, by the time Odiyan has finally taken form out of that dream, the very first seminar there defines a direction strikingly different, it seems, from classical Tibetan Buddhist thought.

I say, "it seems" because for now I am not all that sure. Time, Space, and Knowledge struck out along what might be a new path for Vajrayana Buddhism. Odiyan, in one sense, then might become a new wing on the ever-growing museum/workshop, a timely and needed extension, adding more space but all the same an integral part of the one house, the one continuity.

This is what I thought, and this is what I think.

# Close to My Heart

Ronald E. Purser

Through the years, the TSK vision has been very close to my heart, enduring and deepening in the midst of many personal and professional changes. My contact with TSK began in 1979, when I placed the book on my shelf, but I wasn't quite receptive to giving it my full attention. After I moved to the Bay Area in 1982, I discovered the Nyingma Institute and signed up for a four-week introductory TSK class.

Having a little taste of TSK, the following year I enrolled in what was then known as the "10-month TSK program" at the Institute. Classes were held on Tuesday and Thursday evenings, along with three weekend retreats, on the Space, Time and Knowledge sections respectively. The program was co-taught by Robert Pasternak and Tom Morse. There were only four or five of us in the class, and we worked extensively with the first TSK book. It was by far the most significant learning period of my life. The first book continues to be my favorite, and I consider it to be the foundational or root TSK text.

In 1985, I moved from the Bay Area to Cleveland, Ohio to work on my doctorate at Case Western Reserve University, so for a few years formal study of TSK had to be set aside. However, in 1988, I invited Robert Pasternak to the Department of Organizational Behavior to conduct a weekend workshop for interested doctoral students. Following this, I began going on short TSK retreats at the Nyingma center in Boulder, where Bob Pasternak and Hal Gurish worked extensively with the second and third TSK volumes, *Love of Knowledge* (1987) and *Knowledge of Time and Space* (1990). In 1990, I began teaching at a graduate program at Loyola University in Chicago, and invited Jack Petranker to lead a weekend TSK workshop for graduate students and business executives.

Several years later, in 1994, I offered a graduate-level course on TSK at Loyola University's Center for Organization Development. Around this time an informal TSK 'listserve' was created, and many vigorous discussions ensued between TSK teachers, long-time students, and even occasional contributions to the listserve by Rinpoche himself. Beginning in 1995, I attended several informal summer TSK retreats in Colorado, which I found to be extremely beneficial and inspiring. In later years, when I was unable to attend such retreats due to family commitments, I set aside two weeks each summer for personal TSK retreats. I would usually spend most of the day at a quiet and virtually unoccupied university library, studying TSK inside and going outside to practice under shady trees on the serene campus.

My intellectual involvement with TSK intensified as I considered how the vision made me think differently about core ideas within my academic field of management. In particular, I considered how unexamined assumptions regarding space, time, and knowledge were limiting features in organizational theory. These ideas took form in chapters that I contributed to *Mastery of Mind* and *Light of Knowledge*, books in the Perspectives on TSK series.

In May of 1998, I had the good fortune to attend a rare TSK retreat held at Odiyan, during which Tarthang Tulku met with us for a lively and stimulating seminar. It was also at this time that I met Lee Nichol, a close collaborator of the late quantum physicist David Bohm and the editor for Bohm's posthumous works. Subsequent to our meeting at Odiyan, we began working on a book project that was inspired by our mutual interest in the visionary work of David Bohm and Tarthang Tulku. Starting in about 1999, the Institute invited Lee Nichol to lead workshops and retreats that explored the intersection and dialogue between TSK and Bohm's work in physics. The response has been extremely positive and his seminars are ongoing.

TSK also inspired the creation of the Center for Creative Inquiry (CCI), founded by Jack Petranker in February 2000, and I serve on its Board of Directors. In August 2000, the national Academy of Management—a professional organization composed of some 11,000 professors, doctoral students and corporate managers—held their annual meeting in Toronto and chose as its focus the theme "A New Time." Jack Petranker and I, along with several European scholars known for their academic work on time, participated in a symposium that challenged conventional notions of time in organization and management theory. Barbara Adam and Richard Whipp from the University of Cardiff—both participants in this symposium—subsequently invited me to contribute a chapter to their recently published book, *Making Time* (Oxford University Press, 2002). Drawing upon TSK-inspired insights, my chapter questioned the uncritical acceptance of so-called "real-time" technologies—exposing their contradictions and undesirable consequences. I am continuing to write and make presentations on these themes.

Given the positive response and interest in our ideas, in June 2002, the Center for Creative Inquiry and the Organization Development and Change division of the Academy of Management, which I chaired from 2000–2001, co-sponsored a unique "improvisational conference" in Boston focused on the theme of dynamic time and creative inquiry in organizational change. Papers presented at the conference will be appear in a special issue of the *Journal of Managerial Psychology*. Several participants at the conference described it as a high point of their professional careers. A second conference, on "The Future of Time" is planned for the summer of 2004 at INSEAD Business School near Paris.

Over the last few years, I resumed formal study and practice of TSK at the Nyingma Institute, participating in regular classes co-taught by Tom Morse and Ken McKeon. These classes have focused mainly on the teachings and exercises contained in *Dynamics of Time and Space* (1994). Coinciding with this period, I also found myself being drawn more seriously toward traditional Buddhist studies and methods. TSK has been extremely useful in helping facilitate a deeper understanding of select Buddhist teachings and in preparing the ground and transcending obstacles in formal meditative practice. My own investigations convince me of the point made repeatedly by Tarthang Tulku: that the TSK vision is not Buddhism in disguise. Yet there is also no doubt in my mind that TSK and Buddhism are mutually supportive of each other, and I am grateful to be able to take them both to heart.

# As in a Dream

Piet Hut

*As in a dream, this simple magic*
*Of space and time, of here and there.*
*What is this here, what is this there,*
*What is this magic of location?*
*This ocean of presence, these waves of arising,*
*The wonder of anything happening at all?*
*Time to be open, Space open to be,*
*Living the dream, dreaming our life.*

When I wrote these lines, five years ago, I was on my way from New York to Tokyo, during a fourteen hour flight, in that modern anonymous space in between cultures and countries. I had made that flight at least thirty times in either direction, either in connection with my ongoing astronomy collaborations in Tokyo, or to meet with other Japanese scholars, in philosophy, psychology, comparative religion, and other fields. Yet I always enjoy this no-longer-here, not-yet-there of a long flight, a bit like the space in between breathing in and breathing out, in which the normal flow of events is stilled—leaving room for unexpected openings, orthogonal to the stream of events of daily life.

It was the last day of Obon, the Japanese memorial period in the middle of August, when the spirits of the ancestors come to visit their living relatives for a few days. On this last day, the spirits are guided back to their own realm, sometimes in moving ceremonies such as in Kyoto's Daimonji, where huge bonfires spelling out written characters are lit on the mountains surrounding the city. Like lights lining a landing strip, these bonfires are pointing the way home for the spirits. The last group of fires to be lit, to the west of the city, takes the shape of a gate, as an entrance to the other world.

The next few days, I would stay in the tiny village of Mure Mura, near Nagano, for an informal workshop with three Japanese philosophers, Yoko Arisaka, Nobuo Kazashi and Shin'ya Noe. We would stay in an old summer villa in the woods, to talk about our views of connections between science and religion and philosophy. In preparation for those discussions, I asked myself how I could express my most deeply felt beliefs about the world, about life and death, about reality in all its forms. In other settings, I would have tailored my answer to the specific cultural context, and the specific expectations of the audience. I could have started from my scientific understanding of the world, or from my philosophical views, or from my expectations for intercultural dialogues. But in the setting of a walk in the woods with three friends, all of us eager to open ourselves to a heartfelt form of inquiry, there was no excuse to hide behind constructs.

The first answer to my question came in the form of the poem I started with above. This was the closest I could come, at that time, to express in words what I felt wordlessly: a sense of Being, a sense of presence beyond the dual categories of subject/object, personal/impersonal, free/unfree. I wrote an entry in my computer diary:

> I feel that I am finally beginning to learn to trust and respect Being enough to allow it to shine through my conscious awareness. A Shamanism of the simplest sort: its totems or talismans or sacred objects or tools being space, time, knowledge, and Being. Letting thinking do the thinking, letting seeing do the seeing, etc. In general, let happening do the happening, let arising arise, let appearance do the appearing.
>
> A transformation already. I first thought about sitting back for five minutes of a free sort of meditation. But then I thought to myself: why not stay at the keyboard here, using thoughts, staying in my scholar habitat? As soon as I thought that, looking ahead, I already felt a change, a transformation.
>
> How to describe it, while staying with it? Perception more thin, more sharp, as if everything is drawn with sharp pencil lines. Also a

sense of depth, as if people and things that are twenty feet away seem at oncc much further and distant and at the same time almost within hand reach. A feeling as if I am a hollow tube or flute, with Being flowing through me, both in and as me, and me letting myself be guided by the dance.

There were other things I jotted down, more specifically related to the type of meditative practices I was engaged in, connected with Tibetan Buddhist and Daoist suggestions; including those here would require too much explanation and background. However, I hope that I am conveying at least a whiff of my sense of a 'way of knowing' that I felt very keenly, then as well as now.

So far, this has been a rather personal way to introduce my main topic: Science and Religion as Ways of Knowing. And this presentation is an experiment; I have never before allowed myself to use such a personal tone in a formal conference, for which the proceedings are open to anyone to read. As with every experiment, I am very curious to see the result.

Some may view my introduction as irrelevant, since it is neither scientific nor religious in the strict sense of the word. It is not scientific, in that I do not try to step out of the picture. Nor does it fit clearly within the fold of any one particular religion. Still, I hope that you will appreciate the phenomenological nature of my attempt to help set the scene for a possible dialogue between science and religion.

## 1. STEPPING BACK

Phenomenology, the description of the phenomena under study, plays an important role in almost any way of knowing. Ideally, phenomenology should be independent of theoretical prejudices, but in practice such an ideal is impossible to realize. Rather, we can start to describe what we see happening, aware of our own role in the description. We can then scrutinize that role, to see how far we can outgrow our own prejudices, by noting them and testing them to see to what extent they are appropriate. Normally, we are immersed in both theory and experiment: we live in our ideas and concepts as much as we live amidst all that happens around us. It is a challenge to step back in order to find a 'tidal zone' away from the ocean of pure theory and the firm ground of 'experience-as-we-normally-experience-it'.

The term phenomenology is used in particle physics in just such a way. It indicates a form of theorizing that stays close to the experiments. In contrast to pure theory, such as string theory, aimed at building a comprehensive theory of the material world from first principles, phenomenology does not try to ground itself in fundamental concepts or axioms. It takes a rudimentary form of theoretical framework, within which it tries to construct rough models of the phenomena under consideration. These models can provide guidance for setting up and for interpreting new types of experiments. At the same time, these models act as beacons to guide and test developments on the level of more fundamental theory.

In philosophy, the term phenomenology is used to indicate the approach taken by Husserl and his followers, starting around 1900. Interestingly, in the course of his phenomenological investigations, Husserl developed a method to 'step back' from the phenomena as they present themselves in daily life, including the daily life of a scientist in a lab or behind a desk. In what he called the 'epoché' (a Greek word for 'suspension of judgment'), he advocated a radical switch in attitude from the 'what' to the 'how'. Rather than taking the world for granted while finding our way amidst the many things that populate our world, Husserl invited us to change our focus, in order to investigate how all those things appear. In doing so, he extended William James' radical empiricism.

Through the epoché, all that appears is seen and acknowledged as it appears, in its own structure of appearing. The trick is to refrain from tying down appearance immediately and prematurely to the usual external explanatory framework (of a physical world of objects, in which we possess a body with sense organs and a brain that gives rise to our consciousness that grasps the objects).

At first, it may seem strange to 'put the world on hold', to drop any belief in objective reality as the prior and only 'real' form of reality. But there is nothing magic or special in this shift. Only the result of the shift is remarkable—a form of amazement and wonder. Such reactions can be the touchstone for whether a shift really has been made, or whether an attempt to `put the world on hold' has only been an intellectual game.

Many poets and novelists have testified to such a shift, a dramatic change in experience, away from a belief in a solid world in which we are anchored and towards a completely open experience of the world as bottomless. Several philosophers, too, have given us an inkling of their experience along these lines. The problem is that poets tend to give their experiential report without any theory, while philosophers tend to give only their theoretical reflections while glossing over the experiential component that undoubtedly underpins their theoretical moves towards more open interpretations of reality.

Husserl himself did give ample indications of the fact that for him the epoche was a way of life; towards the end of his life, in his last book, *The Crisis of the European Sciences*, he described it as a "complete personal transformation." While this is an indication of the importance he attached to his shift, the shift itself can be tried out in a surprisingly simple way. Some day, I hope to attend a workshop on science and religion where a few hours are devoted in laboratory mode to collectively exploring such a shift of attention and interpretation.

## 2. BUILDING UP

In science, we filter what we see around us, retaining what can be measured and quantified objectively and leaving out aspects of reality such as beauty and meaning, which do not pass through the filter. The implicit hope felt by many scientists is that ultimately nothing is lost in the process. By initially reducing all of reality to an interplay of atoms and molecules, or further down to an interplay of waves in quantum field theory, or perhaps even to an interplay of strings and branes, they hope to get closer to how reality really is. All the rest then follows from the complexity emerging in the process of building up, from waves to atoms to cells to brains to the human experience that seems to be correlated to our bodies and especially to our brains.

Can reality be understood exhaustively through this approach of building up insight from the (material) bottom up? At present, this is a completely open question. Science forms one of the youngest ways of knowing in human history. While surprisingly effective and productive in its first four centuries, nobody knows how far the scientific way of knowing can be stretched. When we succeed in drawing up a complete wiring diagram of a human brain, together with a detailed functional understanding of how the brain works, will we then understand human experience? It is impossible to say, given that we have just made the first baby steps in this enterprise of mapping and analyzing. Some argue, by extrapolation, that our reductionistic methods will soon 'explain' consciousness. Others argue that such third-person descriptions will never succeed in crossing the gap that separates them from first-person subjective experience.

I expect that neither position will turn out to be correct. In principle, I see no reason that science in its future growth will reach fundamental limits. To the dismay of those historians and philosophers of science who try to show `what science is' and how it got there, it is notoriously difficult to pin down science. Science is what scientists do. And scientists are out to solve problems, in whatever way seems to work.

Like entrepreneurs in business, scientists are extremely opportunistic. Nothing is too sacred for them to to toss it out if there are convincing reasons to do so. Perhaps the most dramatic example occurred seventy-five years ago, with the invention of quantum mechanics. Reluctantly but decisively, strict repeatability of experiments was dropped. What had been one of the corner stones of classical mechanics, the notion that two experiments, carried out under the exact same conditions, would give the exact same outcome, turned out to be incorrect. Instead, nature seemed to show an intrinsic spontaneity undreamt of by previous generations of physicists.

This opportunistic attitude of scientists is their strength. While their views may come across as arrogant and dogmatic, as a result of the frequent overconfidence of individual scientists, collectively science has a built-in mechanism for change. Peer review forms the system of checks and balances that allows for a trade-off between the radical and conservative sides of science. Scientists are extremely radical when it comes to dreaming up new hypotheses, and at the same time they are extremely critical when it comes to testing those hypotheses. In this system, what may look like dogma at any given time is really only a network of working hypotheses, in which each strand is amenable to revision.

Coming back to the question, are there limits to scientific knowledge, I expect the answer to be no. But at the same time, I expect future forms of science to be almost unrecognizable from our present-day vantage point. In particular, I anticipate science five hundred or a thousand years from now will have dropped the reductionistic assumptions that are still largely operational. Already, quantum mechanics has shown us that a purely objective ontology of the world is far more problematic than classical mechanics assumed. How a subject, human or machine, measures an object determines in a fundamental way what the outcome can be. Reality seems to reside as an unformed mixture of actuality and potentiality, until a measurement forces a momentary semblance of actuality to appear. Shocking as this would be for a nineteenth-century physicist, who knows what further developments lie in wait in the next thousand years?

As a specific guess, I anticipate that first-person felt experience and third-person description will both become part of an extended form of scientific method, in a framework that will transcend the current dichotomy. After all, the whole history of the physical sciences is a story of transcending dichotomies. Going beyond electricity and magnetism to electromagnetism in the nineteenth century was a great step, but not as radical as the steps taken in the twentieth century. In relativity theory, going beyond space and time led to the notion of spacetime, where different observers can decompose the same spacetime into different mixtures of space and time, with black holes as an extreme case where space can take on a time-like character. And in quantum mechanics, going beyond being to a description in terms of becoming was even more radical a shift. I wish I could have a glimpse of a preview of the next few radical steps in transcendence in science. That they will occur seems almost certain; what they will be is the question.

## 3. OPENING UP

In contrast to the various disciplines of science that are currently accepted and taught world wide, there is much less agreement and universal acceptance among religions. Even the term 'religion' is highly problematic. It reflects a category in Western thinking that does not allow a good fit for many aspects of, say, Hinduism, Buddhism, Daoism, or many indigenous forms of shamanism. Whether Buddhism, for example, is considered to be mainly a form of religion, or philosophy, or psychology, is a matter of debate. Different answers emerge depending on which aspect of Buddhism one considers, in light of its 2500-year history and its various adaptations and revisions in a variety of completely different cultures.

I have been speaking of science and religion mainly because 'science and religion' is a recognizable handle. Perhaps the title "'Building up' and 'Opening up' as ways of knowing" would be more accurate, reflecting the opposing directionality of the two methodologies. Whereas science presents a bottom-up approach, there is a top-down approach visible at the well-spring of every major religion, in the widest sense of the term 'religion'. And beyond what we normally call religion, this top-down approach can be recognized elsewhere as well. Even amidst what may seem to be the most rational of methodologies, we may encounter the opposing current. Just to mention two European examples: both Socrates and Spinoza combined a love for reason with a love for 'opening up'.

Most accounts of Socrates' adhortation to lead an examined life gloss over the fact that he sometimes was lost in what seemed like a contemplative trance for hours at a time. For Spinoza, too, it is obvious that he was driven by a connection with something beyond the ordinary world of transient phenomena. His way of talking about reality *'sub specie aeternitatis'* (in the light of eternity; literally, as an aspect of eternity), conveys the impression of someone who has found a deep form of peace by opening up to what struck him as a timeless realm.

Distinctions between secular and sacred ways of knowing may be important when assessing sociological, political and historical aspects of religions, and other forms of belief and exploration. And in a dialogue between science and religion, such aspects are important topics to be discussed: ignoring the political and sociological dimensions of either science or religion would present a distorted view of the role they play in modern society. At the same time, there should be room in such a dialogue for a comparison between the personal ways of knowing that underlie both science and religion. What I hope will happen in future dialogues is what I like to call a 'roots not fruits' approach.

## 4. RESPECT FOR ROOTS

It is certainly interesting to compare the conclusions from sciences and religions. For example, we can take the big-bang theory of the origin and early evolution of the universe and compare it with the creation myths in various religions. However, it would be a pity if such comparisons would be limited to the *fruits* of science and religion. It is at least as interesting to compare the *roots* of both approaches. Before asking questions such as "What do science and religion have in common?" or "Are their ways of knowing mutually incompatible?" we may want to ask the prior questions "What is the scientific way of knowing?" and "What is the religious way of knowing?" We often assume we know what science is and what religion is, but do we, really—to the extent needed for a real dialogue?

What can we say about the complementary ways of knowing that I have called 'building up' and 'opening up'? As a starting point, I suggest that we add a form of phenomenology as a third way of knowing, a stepping stone toward crossing the gap that seems to separate the first two. This gives us:

☆ the bottom-up way of knowing implicit in the scientific method, whether labeled as reductionistic or as allowing room for emergent properties to be equally essential;

☆ the level way of knowing of phenomenology, in which we take a step back from our normal involvement with our environment in order to take a better look, without thereby shifting to a different, more fundamental domain 'under' or 'above' our every-day experience;

☆ the top-down way of knowing present in the core of many forms of religious experience, as well as in other whole-hearted engagements with reality that may not be labeled as religious.

Another way to characterize these three approaches is to describe how the world appears to us when we apply them. When viewing the world in a bottom-up way, everything is seen to be made up out of an intricate interplay of matter and energy. When viewing the world in a phenomenological way, everything is seen as a play of various forms of experience. And when opening oneself for a top-down way of letting the world appear, everything is seen in terms of Being, as a play of Being with Itself.

For me, it is a totally open question which of these three ways of knowing is the more fundamental one, or even what it would mean for one to be more fundamental. I can see arguments for any one choice, as well as for a no-choice option, in which none of the three is given pride of place. I certainly understand the appeal behind a `building-up' approach, the challenge to become God-like in our understanding of Nature, or at least as close as we can come to such an understanding on a fundamental level. I equally appreciate the wish for purity that has driven Husserl and others like him throughout history to adopt a radically empirical stance, to come as close as possible to accepting all that appears and only what appears. And finally, I have a deep appreciation for the 'top-down' approach, which for me is as much a way of life as a way of knowing.

The last few years have seen a ground swell of interest in the topic of science and religion. I am delighted and excited to be alive in this period of time, in which we will undoubtedly see a thousand flowers bloom across the vast terrain that we are now entering. Let me just mention here a little flower that I have been cultivating with a group of colleagues: the Kira Institute, which has been organizing annual summer schools for graduate students in the sciences and related areas. How all these different flowers will bloom, and how they will affect future visions and activities of humanity, is impossible to predict. For now, let us just try to be respectful enough to recognize where all these new developments will want to lead.

## POSTSCRIPT

The text above, slightly modified, formed the position paper that I presented as the closing talk at the Future Visions conference in New York City in September 2000, as part of a meeting of the State of the World Forum, which was held there in tandem with the United Nations Millennium Summit. (The conference was sponsored by the John Templeton Foundation, the International Space Science Organization, and the Metanexus Institute on Religion and Science. For more information, search for "Future Visions" at www.metanexus.net.)

Clearly, the text shows the influence on my own thinking of the Time, Space, and Knowledge vision. In that particular paper I listed no references, but in several of my other papers I have referenced the first book in the TSK series, which introduced the vision more than a quarter century ago.

I found the first TSK book in a bookstore in Amsterdam in 1980, soon after it appeared, while I was still a graduate student. That same year I had the opportunity to attend a TSK workshop at the Nyingma Institute in Berkeley, and in following years I attended similar workshops in Berkeley and in Holland. In the '90s I taught a few TSK workshops myself, upon the invitation of the Nyingma Institute.

Back in 1980, as soon as I started to read the first volume, I recognized that this book came closer than anything else I had seen to my life-long inquiry into the nature of reality. Here was a vision based on the theory and practice of a particular culture and background, but stripped of its specific religious rituals and creeds. What was presented was a core of ideas on a par with what underlies the scientific method: starting from a set of working hypotheses together with an experiential method, you plunge in, refining both the experiential method and the working hypotheses while you get deeper into the subject matter.

Many years later, other TSK books began to appear, starting with *Love of Knowledge* in 1987. I enjoyed reading those books as well, and in conjunction with the first book they gave me extra inspiration and suggestions for further exploration. However, for me, at least, the rigor and consistency of the first book were missing in the later volumes, and I'm not sure how much I would have learned from them if I had read them without having had prior access to the first book. This is increasingly the case for the subsequent volumes in the series, and I must admit that I'm rather puzzled by the latest volume, *Sacred Dimensions of Time and Space*.

The combination of the internal logic and consistency of the first book, together with the evocative power of the text, is quite remarkable. It has the flavor and force of poetry, in cutting through conventional ways of looking at the familiar world, and yet it also has the rigor of a scientific or philosophical treatise. Just reading the book attentively, while trying to see and taste each paragraph, gave me an experience similar to reading a book on mathematics: with each statement one is introduced further into a world that is unfolding and that becomes more and more familiar.

While I have been deeply inspired by the TSK vision, I have not yet found a clear way to line up my own evolving insights with those of the first TSK book. I hope and expect that I will be able to do so in the not too distant future, in a more precise way than I have done so far in my more philosophical publications (see e.g. my web site, especially http://www.ids.ias.edu/~piet/act/nat/).

The central notion that I am trying to express in my own writings is the idea of 'freedom from identification'. We tend to identify with all kinds of things. If I say 'I am an astronomer', I point to the fact that I have a job in astronomy, and that I have interest and experience in that area. However, what I really am is not something that can be cataloged so simply. Nor does saying that I am male, that I am a Dutch citizen, etc.

By using our own life as a laboratory, we can explore what it is that we tend to identify with, and with each discovery we make, we can add one more thing to the list of what we have, taking it away from the list of what we think we are. In TSK terms, the path to freedom from identification could be summarized as follows: "appreciate the presence of appearance as a presentation by Being."

# Child's Play

Helena Antonova

During my first contact with TSK, I suddenly realized that the TSK exercises were hauntingly familiar. They reminded me of childhood games I had played.

During the summers in the country in my youth, we lay on the meadow grass, carefree, gazing into the sky without blinking. We tried to embrace the whole sky in one glance without moving our eyes to the left or right. Sometimes we could simply merge with the sky, losing the sense of our body. We were in flight! And so happy, and very still in order not to disturb this fusion with the vast blue space of openness.

Sometimes we imagined what it would be like if we stretched over the whole sky and could look down upon fields and forests, rivers and houses. In our imagination, the body expanded to fill the whole world, and again we were deeply quiet.

Then we imagined that we were living here in this place, but that the time was long ago and we were very young. Or we wondered about the time even before that. What had been there when we were not yet there? When we were not born, what was the world like? When we were not born, were we not anywhere at all?

And what would it be like when we were older, much older and independent? What would we know then? Then, we told one another, we would know how this world is set up, where we had been taken from, and why it was necessary to die. But later, when 'then' was 'now', it turned out very differently. The questions we had cared about so much were swept away as unnecessary, even meaningless. Why ask, we learned, when your questions only bring yearning and anxiety to your heart?

How grateful I am that TSK teaches a way for us to ask once more.

# The Virus against Reality

Meindert Gijzen

*On the seashore of endless worlds children meet.*
*Tempest roams in the pathless sky,*
*ships are wrecked in the trackless water,*
*death is abroad and children play.*
*On the seashore of endless worlds*
*is the great meeting of children.*
— Rabindranath Tagore

In the following passages I have tried to write down some considerations that bear on the questions: What has TSK meant for me up to now? What do I see as special characteristics or unique qualities of this vision? In the struggle to produce these remarks I questioned repeatedly whether I should separate accounts of personal experiences and personal reflections from more general and abstract considerations. I finally decided to let the two coexist, since this is the most accurate rendering of what I have to present.

This does not mean that the present article should be regarded as an ego-document in disguisc. Instead it openly illustrates the principle that nothing can be universal or general without letting it emerge from a personal perspective. To try to hide or brush away this fact would be deceit. Stretching this principle one might arrive at a paradoxical proposition: what is most personal is by nature also most universal.

## The Ship of Adventures

When I was about eight years old I used to play with my nephew and niece at the waterside. We lived near a big lake where, during clear weather, the clouds in the sky would reveal different colors and shapes. One nice day the water was exceptionally low and many pieces of wood and refuse were lying around, exposed to view. The weather was clear but windy and the sky was full with the constantly moving shapes of ragged clouds.

At a certain moment we did not know how to continue our play. While I was looking at a peculiarly shaped piece of wood, an idea spontaneously bubbled up in my mind. On impulse I exclaimed: "This piece of wood is our Ship of Adventures!"

My niece and nephew looked at me, uncomprehending. "What do you mean?"

"With our ship of adventures we can play a game of magic," I said. "All we have to do is look at the clouds. Whenever we change the position of our Ship of Adventures, we will be able to see new shapes and stories in the clouds. We can have lots of adventures! Look!" With that, I changed the position of the piece of wood and pointed at the moving, bubbling and brilliantly white clouds. I began to point out familiar shapes and tell stories about them. "Look, this is so and so, and he is doing such and such. And there is this, and that, they are meeting each other".

My nephew and niece, who were one and three years younger than I was, listened with rapture. I continued telling stories, moving the position of the "ship of adventures" now and then. Our minds were totally immersed in the shapes and forms that we recognized and the stories about them. We were together in the same thing, transported for a timeless moment into an enchanted reality. I believed what I was telling, and even though I knew that adults would put it all aside as nonsense, this reality was valid—as real as anything.

After some time we went home. When we arrived my nephew and niece enthusiastically started to tell their mother, my aunt, about the great adventures we had had. "And who invented all this?" my aunt asked. My nephew and niece looked at me. My aunt stared at me for a moment, in a skeptical and disapproving way. Something inside of me felt awkward, as if I had been caught in a terrible lie. I could only laugh sheepishly. The sensation of that moment, the lingering sense of guilt and foolishness, remains with me even now.

This memory is from a time when my childhood mind was still flexible enough to 'shift realities' at will. Something of that remains with me, a kind of secret opposition to the idea that our ordinary daily world is the 'one and only' truth. To be able to function in an acceptable way, one has to put up with this proposition, but I was exhilarated in later years to find that my opposition was shared by others. Shamans and the masters of religious traditions all over the world and in all times have pointed out that reality can be experienced as flexible, open as a dream, and that in fact keeping up the idea of the "one and only" solid world that we seem to live in itself requires a certain kind of well-conditioned imagination.

Through my readings and studies I discovered that we human beings all seem to use the imagination to create reality all the time. Tarthang Tulku describes the power of imagination very poignantly (VOK, p. 8):

> The patterns we embody in our action and our being gain power over us because we 'know' them to be true. But what is the source of that knowing? It seems we accept the way things are because we imagine that they cannot be otherwise. The lack of vision we experience—even the impotence of insight to affect change—indicates this hidden power of imagination in operation. Can we imaginatively embody this insight?"

According to an American Indian teacher, the author of several popular books, we all live in a collective dream, which he calls "the dream of the planet." The problem is what he calls our "domestication." Sooner or later in the course of our upbringing, our minds will be shaped and conditioned to perceive reality in a highly pre-programmed way. The open and unstructured mode of viewing the world that seems to be available to us as children slowly subsides, replaced by a way of perceiving and using the mind that is governed by labels, routines, stories and very large blind spots. We start to think of ourselves as real persons with a separate identity. Gradually the labels and structures, the stories that we have adopted, take over our lives. At the age when we enter society and envision a certain career or role ahead of us—however great or humble it may be—our so-called 'real life' starts.

Many books are written out of homesickness and nostalgia for childhood, in which the petty restrictions of adult life did not yet apply. Yet there seems to be no path leading back to the freshness and openness that we knew as children. Instead, we have to cope with 'the real world'. We may take refuge in the apparent safety of a job or career, a relationship or family life. If we feel that we cannot face up to the demands that society puts upon us, or if we feel that life is too painful or threatening to face, we may even take refuge in an addiction or negatively deviant lifestyle. Whether we choose a socially acceptable route or get trapped by a more deviant one, we are more or less completely ensnared by the circumstances and histories that we have generated. The perspective of freedom seems to become ever more faint and unreal.

So what is freedom, and how can it be regained? And how to unravel this mystery of reality? How are freedom and reality connected? Why did we have to invent a fixed view of reality at all? These questions have been nagging at me since I was forced to leave childhood behind.

## Encountering Time, Space and Knowledge

As a Westerner with an academic background, I used to have great difficulty getting in tune with the more formal religious aspects of most teachings I studied. In the course of my education as a child and youngster I had become alienated from my Christian background, and apart from the fascinating stories in the Bible, I had no positive experiences with the idea of a personal God. If anything, my relation with Him was quite disturbed. At the same time, the loss of several family members imprinted on my mind the reality and inevitability of death at an early age, and forced me to start looking for 'solutions' to the problem of life and death. While still quite young, I had embarked upon a journey from which there is no returning possible.

Against this background, my first encounter with anything related to the TSK vision came when I stumbled upon a copy of *Love of Knowledge* in a bookstore in The Hague, about a year after its first appearance. I was immensely intrigued by the subject matter, as well as the many pictures and diagrams, which seemed to jump from the pages. The decision to buy the book was made instantly and intuitively.

This was a book that fascinated me from the outset, not because the message it radiated was so profound and clear, but because it seemed to be entirely devoted to the Riddle of Reality. In those days I was blessed with having a large and lusciously green garden at my disposal. One of my favorite pastimes became hiding away in the back of this garden with my copy of *Love of*

*Knowledge*. I still have clear memories of sitting there, amidst the fragrance of flowers and blooming shrubs, the singing of birds and the buzzing of insects.

I did not read the book in a conventional, linear way. Rather, I sniffed the pages, reading some lines now and then, as if they were a kind of poetry. In the spaces between reading one passage and the next, I rested, observing nature as well as my own thoughts.

The effect upon me was strange. Although I did not rationally understand what the text was about, I nevertheless received inklings and intuitions about what it might refer to. There was a slight alteration in my usual perception of reality, a subtle but pleasant confusion that lingered even after I stopped reading. It was as if I had received a premonition of a strange and unknown vision of reality; as if I could smell it without understanding anything about it. Sometimes after reading, I would go for a walk at dusk, still infected by a quiet sense of wonderment. A strange and peaceful sense of freedom filled me. I remember once the sudden and silent appearance of an owl gliding overhead, as silent and unexpected as Knowledge itself. . . .

The search that ensued would eventually lead me to the study of the TSK vision, either with other people in workshops and courses, or on my own, through reading and working with the exercises.

## Inner Technologies of Freedom

A little bit of inquiry reveals that the concepts of freedom and reality have a common background. Both rely upon an immense field of hidden meanings and silent assumptions, culturally, ideologically and personally conditioned, that govern our view and experience. It is easy to loose sight of the fact that any description of reality can be nothing more than the creation of a map. The territory, the 'real thing', is unknowable, except as direct, momentary experience, and even then it is open to questioning. Yet the descriptions of reality we generate have immediate implications for our understanding of freedom.

For example, the materialistic and mechanistic worldview that dominated science throughout the first half of the twentieth century describes human beings as intelligent apes, hopelessly at the mercy of biological drives and instincts. The corresponding conception of freedom could extend only into the political and practical dimensions of freedom, leaving the inner dimension that freedom ultimately points to largely out of sight. This limiting conception still exercises enormous power, so that modern society regularly invests its hopes for a better life and true happiness in better material circumstances, more advanced and systematic research, more refined technology, new forms of behavioral control and psychotherapy, new medicines and drugs, and the unbounded proliferation of amusements geared to mass culture.

The history of philosophy and religion offers countless variations on a rather different approach. These alternatives view man as trapped in a web of meanings, a prisoner of structures imposed not only by society and biology, but also by his own mind. This view of reality produces a large number of what could be called Inner Technologies of Freedom ("ITF" for short), whose aim is to liberate humanity from the bondage of ordinary life and its suffering. These ITFs share the following general structure:

1) A chart or map describing the supposedly true laws according to which reality and the human mind function. The hidden structure of reality and man are thus laid bare and explained. We can call this *the explanation*.

2) A set of instructions, rules, exercises or practices that need to be applied to develop the full potential of human being and be liberated from all of this. We could call this part *the prescription*.

3) An end result or ultimate state (e.g., fully developed Man, the Kingdom of Heaven, Enlightenment etc.). This part could be called *the promise*.

In current Western society many different ITFs vie with each other for popularity and validity. But almost all of them fall victim to a universal trend that manifests again and again in history. As more and more people devote themselves to the practice or implementation of a specific ITF, issues of social power and hierarchy become more and more important. At the same time, the stories, writings and occurrences connected with these teachings gain weight and importance in themselves, providing a background of myth and meaning and a fixed framework for reference, into which any new incidents or anomalies are absorbed and digested.

In other words, the ITF, which aimed originally at liberating man from his metaphorical prison, evolves via its background of meanings, stories and explanations into yet another source of conflicts, social strife, and misunderstandings. It continues and even exacerbates the problems and circumstances it was originally supposed to solve and transform. A new prison is created from the many meanings, assumptions, prescriptions and injunctions that the ITF generates and collects around it. In the process the perspective of freedom becomes ever more distant and dim. The human impulse toward altruism and religion undergoes a slow and hardly perceptible deterioration towards ideology and power politics. It would be superfluous to mention real-life examples; the newspapers and news reports testify to all this daily.

## Reality as a Secondhand Affair

The notion that this ordinary day-to-day world that we experience and that we have agreed upon to be this and that way is not what it seems to be is not new. It can be found in many philosophical or religious systems and ITFs. In the common way of viewing the world, reality is out there, an objective and substantial world that functions according to consistent laws that cannot be denied. What is in here, the so-called subjective realm, is much more elusive and unreliable. The tree that I am looking at will not disappear if I close my eyes or evaporate if I direct my attention elsewhere, nor will it allow me to walk through it. On the other hand, the inner image of the tree or the thought about it may change or disappear as soon as another thought arises. On closer analysis, however, this ready distinction becomes questionable, since in the end the subjective and the objective components of experience cannot be separated.

In the world of social interactions the distinction is even more unreliable. Stories, labels, and former experiences seem to determine the texture, shape, and course of our present experience almost completely. Interpretations, reputations, histories, moods, power structures, emotional colorings, etc. all play a crucial role in determining the direction, shape and taste of our experiences. Reality in the social context depends totally on what people have agreed upon. Our lives seem to unfold in the context of a second-hand world of labels, stories, stories about stories, relational patterns, etc. The outer social network and context—the roles we play, the established patterns—are all crucial for another, inner network formed of events and psychological patterns. This inner network constitutes the structure of ego or self that has formed out of the many events and psychological patterns from the past.

Looking at how these paired networks of restrictive meanings are inseparably intertwined and connected, we may wonder: where is freedom to be found? TSK explores these matters in far greater depth than this brief analysis is able to reveal. The TSK vision and practice of the exercises (in my own experience as well as those of others that I know) present themselves as unusually powerful tools for opening up restrictive patterns, inviting awareness, and allowing transformation.

## TSK: Vision or Virus?

The first volume in the TSK series was subtitled "A New Vision of Reality". The author made it clear that this new vision should not be regarded as a disguised way of presenting Buddhist ideas, nor as a new kind of philosophy or religion. Instead, the TSK vision presents a way of questioning and inquiry into the very heart or essence of reality. The process of discovery it opens does not seem to lead to any new view of a more (or less) solid structure underlying reality. In terms of the maps I spoke of earlier, in discussing the characteristics of ITFs in general, we could say that TSK seems to question and challenge all fixed maps of reality without offering us a new ('better') map to rely on. As Tarthang Tulku puts it (TSK, p. 200):

> The issues which have just been presented amount to a challenging of what ordinarily seems unchallengeable. On the other hand, the challenge is not itself grounded on something else being claimed to be more real or true.

In this sense, TSK does not offer a new vision at all, at least in the sense of a description or explanation about the 'true' state of affairs. What it does offer is a new methodology for opening up awareness, perception, and reality. For if everything is ultimately nothing other than Time, Space and Knowledge, and these three are totally open and unbounded, offering no reference points, what remains?

Many people I have known who met and studied the TSK vision have had a problem with this openness.

For them, TSK may be liberating, but it is also bewildering and even somewhat frightening. I think this danger lurks most for those that approach TSK too much from an exclusively intellectual viewpoint. As soon as one starts practicing the exercises, one finds that this openness—as far as it is an immediate experiential reality—can be deeply interesting and exhilarating. One discovers that Knowledge provides a nourishment more satisfying and subtle than what one can get from an intellectual approach.

Some people start the practice of TSK explicitly longing for or expecting inner change. They realize that their usual ways of functioning are not satisfying, or they are disappointed with the way their lives unfold. Initially this motivation may work well, for the practice of certain TSK exercises may bring strong momentary shifts in one's ways of experiencing. Later, however, the desire for inner changes or special experiences can lead to frustration. As expected or desired effects fail to appear, practicing the TSK exercises can even become boring. The ego may fabricate a story about how much it has already changed, what insights and special qualities it has developed, how it is going to change in the future, or even how it will eventually reach The Ultimate Goal (enlightenment, or whatever). In TSK practice, changes do arise sooner or later, but they may appear in disguise. They may also not be what one expects or wants to happen at all.

There is another way to look at this. Imagine that human beings, like computers, have an operating system, one that makes possible our normal way of functioning in the world and perceiving reality. The human operating system may run smoothly with respect to producing certain results (such as power games, strife and emotional dramas), but it slows down, freezes, or crashes completely when one tries to cultivate qualities such as compassion, endurance, wisdom, and so forth. Our conditioning and upbringing, as well as any ITFs that we have incorporated, function like more or less stable software that runs under the operating system, but only for specific tasks and within a narrow range of possibilities. As soon as we try to change existing patterns or to activate new possibilities, problems arise.

In this analogy, the practice of TSK (whether through inquiry by way of questioning, practice of the exercises, or ideally both) could be thought of as akin to a computer virus, a tiny program designed to infiltrate and disrupt or disorder the computer's operating system. Like a computer virus, TSK seems to disorder the patterns of our mind in subtle, sometimes hidden, but usually penetrating ways. But the effects are wholly positive. Patterns and tendencies seem to lose their restrictive and compulsive quality, becoming more or less transparent. The result is to clear a space in which the potential for new possibilities is available.

Many people who practice TSK (or any other spiritual discipline) will feel that change comes too slowly, or not at all. But I would reverse that logic. In my view, it is impossible not to be affected by the practice of TSK. The TSK virus disorders the narrow and restricted patterns of the mind in ways we may not be prepared to recognize. In particular, it subtly affects the ego as the main operating system of the mind.

An essential characteristic of the TSK virus is the fact that newly formed routines and patterns will, in time, also be newly affected and disordered. Just as computer viruses can replicate endlessly, so the TSK virus opens the door to perpetual renewal. In fact, the TSK virus will even disorder itself. This will happen whenever "freezing" tendencies start to appear. For this reason, TSK cannot be compared to traditional ITFs. It seems to offer the possibility of liberating human beings from the bondage of all inner restrictions without inviting potential drawbacks.

Once you start studying the TSK vision, the influence of the TSK virus seems to both irresistible and irreversible. To vary the image, it is as if TSK injects the nervous system with antibodies against rigid or dogmatic ways of thinking and acting or deceptively attractive doctrines. So here is a warning! If you want to keep on functioning in limited and repetitious ways, want to keep up your views of the world and yourself, and long for a safe and fixed perspective on the way things are, keep away from the virus, and take precautions to operate only in a sterile environment.

## Advantages of TSK Practice

We might ask: What is (or could be) the surplus value of the TSK vision in a world that is already saturated with teachings, schools of thought, and practices of religious, psychological, spiritual or other origin? Many people are perfectly happy and satisfied with the particular system, worldview, teaching, or belief that they adhere to. And those who are dissatisfied or thirst for something new have ample choice in today's gigantic supermarket of spiritual and religious disciplines.

One thing is for sure: the pile of TSK books produced till now, combined with their abstract contents and the complexity of the language they use, will not be particularly inviting or encouraging for new readers, with the possible exception of those who simply love to be engulfed by abstract matters. Why then take the trouble of reading these books and practicing these sometimes arduous, boring, or difficult exercises? In the course of the years that I have been doing so (with differing degrees of devotion and intensity), some answers have come up. They are of course personal, but I think they may also have a wider applicability.

- ☆TSK practice seems to discourage—or at least to be indifferent to—any kind of religious or political fanaticism or the creation of any kind of group identity. It is very hard, probably impossible, to create and distill an ideology based upon unrestricted openness and inquiry. As long as true inquiry and questioning abound, superficial social hierarchies or power structures will inevitably be undermined and disrupted.

- ☆TSK is totally compatible with the most fundamental and noble principles underlying science and democracy. The quest towards truth, equality, sharing the fruit of inquiry, dialectical exchange, curiosity, etc. can all be found in TSK.

☆TSK seems to penetrate to the core of what we call reality. Instead of creating distractions that divert us from our immediate experience, TSK actually directs our attention to the whole of our present experience, thus revealing any 'blank spots' or neglected areas.

☆TSK does not create a distinction between the psychological and the spiritual levels of the human mind. Unlike many systems or teachings that take this distinction as their starting point, TSK lets us investigate these realms in their original unity.

☆TSK promotes a healthy attitude of being open to change and working with changing circumstances. It lets us be 'on guard' and critical in a positive and constructive way. Through developing such an attitude, we can not only adapt to, but also tune in to and actually make use of new developments.

☆The practice of TSK lets us connect directly to the healing capacities available within the human mind and organism as a whole. A good example of a practical application here is in working with pain, both psychological and physical in origin.

☆Although TSK promotes investigation and penetrating inquiry, it respects whatever belief systems or religious convictions people may have. It can be practiced along with any kind of religion or other spiritual orientation, for it makes no value judgments as to what is better. Practicing TSK, we are thrown back on our own critical faculties and intelligence to make distinctions or decisions.

☆TSK encourages a playful attitude towards psychological problems or mental patterns that we usually consider to be 'serious business'. It thus promotes lightness, new perspectives, and the possibility of significant change in situations that otherwise seem stuck or frozen.

## Conclusion

The foregoing is not meant to interpret the depth and profundity of the TSK vision as a whole. It is nothing more than an account of some personal experiences and reflections. If I should draw a conclusion at all, it is that I see TSK as holding great possibilities and promises for the future. Adapting the quotation from Rabindranth Tagore at the beginning of this article, we could say:

*In the open space of endless worlds*
*TSK students meet.*
*Economics and politics rule*
*the minds of human beings,*
*War is never far off, death is near at hand,*
*and TSK students practice.*
*In the open space of endless worlds*
*is the great meeting of TSK students.*

# Celebrating TSK

Ruth Fellhauer

*Those of us who understand that TSK can come alive in our experience have a responsibility to develop knowledge and share it with others.* — Tarthang Tulku

The above comment by Tarthang Tulku in a TSK newsletter encourages me to write this brief response. May these words be accepted as an offering of appreciation, and may they inspire others to offer more skillful and appropriate responses.

For me, the TSK series, and especially *Dynamics of Time and Space*, have been and continue to be a constant source of surprise and delight. The pages seem filled with liveliness and light, playful and serious all at the same time.

Why do the words seem to move around on the pages . . . or fly away to remote places on other worlds, at other times? Let's visit Kepler and look over his shoulder as he writes his *Harmonies of the World*, listening to the celestial music of the stars. Moving on with time, let's discover Einstein experimenting with pure thought, thinking the impossible, "running with light." Everything in transit: words into musical notes, or poetic forms into mathematical equations.

If we ask what kind of magic this is, the chapter in DTS called "Returning to the Nameless" answers:

> The power that enables words and images to establish the forms of our reality is the power of the communiqué. At times when we access this power directly, we understand that what is identified as real is more accurately the outcome of an act of projecting: that 'reality' is the translation of words, concepts, and images into another dimension."

To do this 'translation', students of TSK may wish to deepen enjoyment and clarify insights by exploring the power and beauty of words themselves. The language used in all the books of TSK Series provides a fine opportunity to improve communication skills. It is surprising how much this Tibetan author can teach us about our own language.

The reader will often come across the term 'field' in these books. It almost serves as a series trademark. The word carries a provocative energy, rich with suggestions of space, expansion, and play. Scientists have their field theory and a farmer tills the field. There is a football field, a Buddha field, a Pure Land, an Elysian field; an open land; a place of battle or campaign; an area in which a force is effective; a field of view; or a field of endeavor. TSK can certainly claim coverage of a vast field, a field of fields. Getting a feel for the field may help us play the game.

## Turning Darkness into Light

*In this pure world, there is no sense of loss and of complicity in loss. No first or second thought arises to comment or project subjectivity. There is no one who conceptualizes or judges, no one to victimize. Although there are meanings, they never depart from clarity. The mind is not preoccupied, and there is no being concerned with past or future.* — DTS

A physicist[1] writing of light and mind states, "It is ourselves whom we study in studying light." And, "knowing is seeing knowledge as event." The skylight of mind opening to space-light and time.

I gaze at a map of the Orion Nebula: astronomers call this a birth-space of stars, a nursery for very young stars. Tonight while I gaze, the spacecraft Galileo is circling Jupiter, the king of planets, to become that planet's umpteenth moon. Knowledge of this event—does it serve to hold our planet together? The physicist again: "Seeing light is a metaphor for seeing the invisible in the visible, for detecting the fragile imaginal garment that holds our planet and all existence together."

Dancing together on this blue planet Earth, "where accurate saying falls away," we movc the outside dance to the private ballroom inside, allowing the strong arms of knowledge to lead in the rhythm of a music without words. The dancers who enter that special place are utterly free to come and go as they please. "It is an experience both specific and immediate: a loosening in the body, in the head, in the mind—a lightness let loose everywhere." Let's dance!

The message of TSK comes from a practicing magician giving away his professional secrets and enjoying every minute of it.

> It is no ordinary magic show that knowledge presents. The magic of knowledge is the magic of our own being. It is the cosmic show of space: the changes of the seasons, the dance of the cosmos, the times of our lives. As a part of the presentation, we can engage with knowledge intimately—engage as lovers are engaged.

Or again, "It is the performing of the play for the sheer joy of it."

## NOTES

1. Arthur Zajonc (1993). *Catching the Light: The Entwined History of Light and Mind.* (New York: Oxford University Press). Quotes without other attribution are from DTS.

# Light and Good Fortune

Ken McKeon

On the cover of *Dynamics of Time and Space* is a stunning fractal image. The image presents a vision of jeweled clarity, a brilliant sky of light that meets the horizon of a mirroring sea. The sea's textured form is blurred, yet its depths are above in the open and radiant sky. It is beside the point, perhaps, to try to make specific meaning out of the image, for the stress is on the evocative and symbolic, not the reductive and ordinary. In the beholder of the image there is a felt presencing of knowledge and good fortune. The knowledge is signified by light and the good fortune by the promising familiarity one feels as inner knowledge. It is these twin signs of welcome, of light and good fortune, that signal doorways to the vision of time, space, and knowledge.

*Dynamics of Time and Space*, like the earlier volumes in the TSK series, is a mix of theory, practice, and commentary. The theory is presented in the familiar tripartite structure of Space, Time, and Knowledge. Again as in the previous volumes, the theoretical presentations are rational and beckoning. Tulku insists on clear language even as he stretches the conventional inclinations of language. The meanings of words mutate on the page of the text and in the mind of the reader. The term 'conduct', for instance, speaks to the ways in which we manage our lives, but it also puns out in precise triples of conducting music, leading for ward, or transmitting time much as a wire may transmit electricity. The transmission of this visionary inquiry is rational, beckoning, and playful.

A hallmark of the TSK series is the stripping of old forms to present an account of and an approach to being in clear, denotative terms. To some readers this has meant a deadening of the visionary import and relevance of venerable traditions. Tarthang Tulku has found a way to eat cake and have it too. He decries obfuscation yet incorporates the past at the same time. The free movement of language and mind over time is, perhaps, the signature of a true master of knowledge.

Knowledge is the honored term of the vision and the honored term here as it does its wonders. Consider the following passage (DTS, p. 159):

> Inward knowledge cuts through the mists of mystification. . . . It rekindles the light of the phenomenal within phenomena. It offers the mystical as wholly available, rejecting incomprehensibility and ineffability for the sake of knowledgeability.

In the above passage, knowledge turns inward to "cut through mists," "rekindle light," and make the hidden available. The metaphors of sword and fire are resonant and familiar but Tulku is doing more than substituting new terms for old. He is presenting a direct path to ourselves and being. The reasoning mind can trace through the philosophic pathways Tulku presents. The world of the writer is knowable and mysterious at one and the same time.

A central paradox of the vision is that mystery itself can inform the truth of experience. Mystery is the evocative presencing of time. Mystery is a central and stimulating feature of vision that does not close down, does not deaden, does not fix or hold. The visionary impulse of *Dynamics* is not to reject one ordering or conceptualizing of the real in favor of another. To the contrary, the visionary impulse is to open and see the inner workings of every order, to penetrate and reveal the dynamics involved in the appearing of time and space, the multitudinous blooming, as it were, in this infinitely fecund garden.

The grip of the past, whether it be the past of a moment, of a habit, or of an historic movement or belief, can be opened, can be seen through until the holding claims of truth, of likely trend, or even fate can be held to, loosened, dismissed, or reformulated, depending upon the desired play or (in other words) the appropriate choice we make as we conduct our lives.

The thought of human beings having choices to make in their lives may strike us as both liberating and perhaps illusionary. We may well try to do or be this or that and find ourselves ending up as pretty much the same creature. It's almost as if we engage in projects of development and at the same time hold fast to the expectation that our efforts will finally be futile, or are presently and habitually silly. Given Tarthang Tulku's stress on the persistence and insistence of any past form on perpetuating itself, such gloomy insights are not surprising. Yet the last section of *Dynamics*, called "Transformations," offers a counter. It extends to the reader ways to work directly with embodied presence in and as time and space.

The "Transformations" section is comprised of practices and commentaries grouped according to theme, power, and intent. Although a section in itself, the presentations of Transformation can be seen as aligning with the theoretical sections of the text. Those engaged more in practice than reflecting may well jump into this section head first, seeing in it the possibility of health, freedom, and meaning.

The practices of *Dynamics* are well worth exploring, worth trying out and trying on. They offer ways of direct exploration and transformation. In working with them one may well find aspects of light and good fortune, those signals mentioned earlier. The practices call for honesty and courage; we begin with who we are. And the practices offer the most thorough introduction and invitation to light and the quality of illumination that I have found yet in the TSK texts.

We may gain from these practices the embodied realization of who and what we are becoming. There might be choices along the way. More effective, enlivening, and satisfying knowledge to guide those choices would surely be of help.

My sense is that in *Dynamics of Time and Space* Tarthang Tulku is offering to the students of TSK a field guide of sorts, suggesting paths of fruitful inquiry. The result is a significant extension and deepening of the TSK vision. The book's intellectual clarity and significant promise may well serve to stir and encourage those who are finding themselves within the vision.

# Prior to Theory or Practice

Lee Nichol

Since I have been exploring the TSK vision, I have found myself engaged with a recurrent dichotomy: the distinction between theory and practice. In one way this dichotomy is quite practical: "Now we will discuss things a bit, and later we will do some practice." But there is a further sense in which preferences are expressed: "Enough theory—it's too abstract. Let's do some practice." Or from another angle, "I would really like to discuss this more fully and understand the meaning."

Then there is the common-sense view in which theory and practice are integrated—understanding theory can deepen practice, and practice can inform theory. For students of TSK, this perspective raises another issue, often expressed as, "How can I integrate TSK into my daily life?"

At one point some of these issues arose in a TSK online forum where the most recent of the TSK books, *Sacred Dimensions of Time and Space*, was being discussed. One posting noted the absence of exercises in the book and mused as to whether exercises would be forthcoming. Subsequent postings suggested creating exercises to go along with the book. When I first read this suggestion, one part of me started thinking of exercises right away.

But the general orientation of *Sacred Dimensions* has encouraged me to reconsider this issue of theory, practice, and their integration. To the extent that I understand it, *Sacred Dimensions* seems to be an exploration founded on the (apparent) irreducibility of the experience, "I am here." I find myself experimenting regularly with this little conundrum, as well as variations such as "something is happening" or "there is consciousness."

I find this experimentation uniquely valuable. It seems to reveal experience as experience. If I am anxious about an upcoming meeting, "I am here" can reveal the movement of anxiety as experience—more fundamental than what we call "anxiety." If I am delightfully immersed in the quality of the Hopi pot that sits next to my desk, "I am here" can reveal the movement of delight as experience—more fundamental than what we call "delight." Repeated experiments point to the astonishing fact that experience exists at all. There could just be nothing. But there is not nothing. Something is happening!

Then, of course, comes the freezing aspect of "I am here:" I am here, the pot is there, and it is mine! Or: I am here, theory and practice are there, and I prefer one, the other, or both. In this way, a world is spun into existence, with me at its center. In part, *Sacred Dimensions* seems to be pointing to the nature and structure of such a "world"—our experience—as it unfolds in this way.

But *Sacred Dimensions* also encourages us to explore experience quite differently. We are urged to inquire into the 'prior' of given experience, the 'prior' of me and my world, and thus necessarily into the 'prior' of theory, practice, and their integration. *Sacred Dimensions* suggests that we may have access to genuinely fundamental aspects of existence, prior to being thrust into the kaleidoscopic seduction of here and there, this and that, like and dislike.

Now, is this very exploration theory? Is it practice? Is it both? Whatever we claim, we immediately establish references points, all of which are equally downstream from experience as experience. In this way we fragment the naïve vigor of our inquiry. But if we explore the suggestion that our given world has a 'prior', we can consider that both theory and practice arise from something more fundamental than themselves. Perhaps, in Tarthang Tulku's terms, they are children and have parents.

Or, to use other TSK terms, perhaps theory and practice are read-outs. In any moment of 'doing' practice, or in any moment of 'conceiving' theory, can we sense the torqueing of experience that gives rise to these image-forms, which first become fixed and then become preferences? Can we sense directly the coming into being of what is eventually called theory? Can we sense directly the coming into being of what is eventually called practice? Equally, can we sense the coming into being of what may be called the integration of the two, including the intention "to apply TSK to the rest of my life?"

Whichever of these questions we might choose to examine, Sacred Dimensions keeps pointing to the 'prior'—a pregnancy of being that allows anything at all to appear, including these various read-outs. For me, considering this quality is both interesting and provocative. But in another, perhaps less abstract sense, this quality can seem boring, useless, or fanciful. One part of me says to another part, "Don't be stupid. This 'prior' is all imagination. You have more important things to apply yourself to. Get back to something that makes some concrete sense."

At such a juncture, I can't help but wonder: Am I pulling away from this odd borderline of 'prior' because I can't quite say what it is? Because I can't turn it into an object that can be inspected?

What if the 'prior' is what is allowing me to look at all? What if it is utterly alive, palpable—allowing blood to flow and stars to shine—but can never be the content of my experience, because it is not possible to get that kind of distance from it? It seems this would make it deeply important. But 'in practice', I am then inclined to dismiss it. It's not something I can 'work on'. I can't really discuss it. It has no entertainment value. Yet, in spite of all that, 'it' actually seems to be present, like a ghost.

The very point of both theory and practice in TSK may be to help us develop a sense of fluid intimacy with this 'prior'. With care and lightness, perhaps they can. Yet it is so easy to be deflected by a directional/result-oriented mentality, a measuring mentality, a 'get it' mentality. If we remain wary of such traps, "I am here" and its concomitant 'prior' may serve as potent clues, helping us avoid the irony of using theory and practice to subtly confirm a world of frozen objects and preferred experiences.

# TSK and a Scholar's Personal Journey

Charles T. Davis, III

Until my late thirties, I was obsessed with the fear of death—religious study and psychoanalysis notwithstanding. In a seminary class on Christian missions, I read the *Bhagavad Gita* for the first time and found it to be a totally boring book. My response puzzled me, because the *Gita* is widely recognized as one of the world's great scriptures. What did others see that I was missing? When I read Juan Mascarŏ's (1978) translation, I found out. I wept with Arjuna and thrilled to Krishna's counsel. In a moment of ecstasy, my experience was transformed in some indescribable way. New vistas of existence opened, and my life-long fear of death simply dissolved.

A short time later, I was in my office at the university when a young woman came to talk with me. She began with language that seemed almost surreal. "Since you are so wise," she said, "I have come to talk with you about my problem." Nothing could have made me feel less wise at that moment. "Alright," I replied very hesitantly. The young woman then informed me that she had learned that she was dying of a cancer. She, a Methodist, wanted to talk with me, a Methodist, about Life and Death. "How," she asked, "could she go home in two weeks for Easter and tell her family?" The contrast between the celebration of Life and her own devastation imposed a heavy burden upon her. I remembered my experience of the Gita and knew that there was no way that I could impart that experience to her. I longed to the core of my existence to transfer my experience of escaping the burden of the fear of death. It seemed so unfair that I had the experience when she needed it far more immediately than I. She went on to tell me how alienated she felt. It was as if her warm consciousness were suddenly trapped in a cold, untouchable body. The only comfort I could offer was by placing my arm around her in the hope of conveying to her that she was not rendered alien and untouchable by the diagnosis of cancer.

A few nights later, I had a dream/vision. A voice directed my attention to a beautiful field of blue; as I watched a small pool of golden light emerged from below; it formed a single petal of the golden flower, as shown in the figure on the previous page. I was to imagine myself tubing down the river of light and to experience the great joy of the ride. After I had experienced this, the voice advised me that the going out from the center is life and the ride back to the center is death: two stages in one joyous journey. I was to share this vision with the young woman for whom my experience of the *Gita* would be inaccessible.

Then, petal by petal, the Golden Flower formed (above) and progressed to the image of the Golden Sphere. The visionary experience ended with the admonition that I was to search my own tradition and confirm that the vision was ageless and shared rather than being a personal delusion or compensatory fantasy.

I looked forward to the next visit scheduled by the student so that I could communicate the vision of the Golden Flower. She never came, nor did she return to class as far as I could determine. When I graded the final exam, I was startled to find a paper with her name upon it. I was left with an eerie feeling. How could I have missed seeing her! The young woman now seemed more like a heavenly visitor than a real person. I came to feel that perhaps it was not she who needed the vision but myself. I have never doubted that it was the deep sense of empathy and compassion that I felt for her plight that opened my sight. The vision was, in a sense, her gift to me.

In the Fall term, I shared my story with a class on advanced Christian thought and asked them to keep their eyes open for possible analogies. I was overwhelmed by the response. The very next class, the students returned with a number of examples, and the process continued throughout the term. One student even mailed me a sand dollar from the beach with comments about the conformity of the Golden Flower with nature—a fact that can be confirmed by examining snowflakes as well as flowers.

It was my good fortune not long after to receive an pamphlet from Dharma Publishing promoting the first TSK book. Tarthang Tulku's *Time, Space, and Knowledge* became a constant companion and an aid as I endeavored to discover how the Golden Flower, or the Three-Dimensional Cross, might be similar to or different from the cross formed by the interaction of Great Space, Great Time, and Great Knowledge. In my experience, the vision of TSK and the Great Flower were mutually confirming. I came to think of them as being like the focal points of an ellipse. Both were needed to generate the experience of freedom for me. I needed the direct experience of the impact of compassion upon knowledge in order that I might experience the truth of the heart. I also needed a confirming voice from outside—from tradition, even from outside my religious tradition—lest I mistake the vision for delusion, the private reality of the mentally disturbed.

Over a period of years, the vision spontaneously renewed itself, showing me additional configurations. First, there was the awareness that each life is a flower. Formed in the image of the whole, each individual life is attached to the end of a petal of the Golden Flower.

All that passes through the central flower, which for me is also the Mind of Christ, passes through the mind of each person who is open to the Spirit that is given.

I also saw that my own Protestant tradition tends to see only a part of the Golden Flower—the center and the terminal points, without any sense of the flow of the Spirit that forms the connecting petal (below).

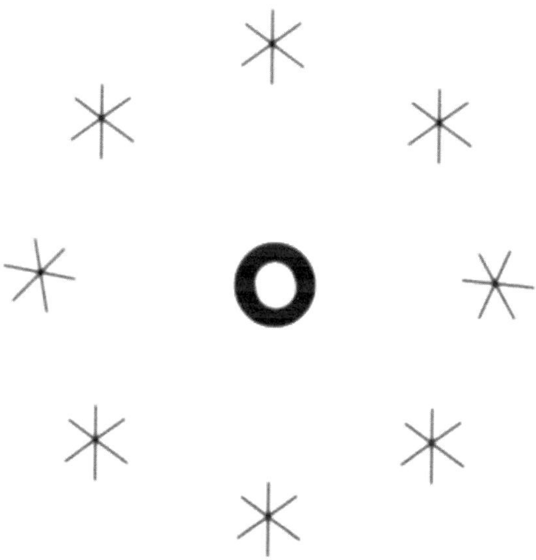

Sin is thus believed to have created an unbridgeable, ugly ditch that separates mankind from a morally distant God.

The vision also indicated the plight of those who affirm with modern science that "reality is a function of matter." From this viewpoint, one sees only the isolated persons; all vision of the larger petals and the center is lost (below). Human beings appear to be isolated in the vastness of the cosmos like billiard balls upon a table.

In a world where one "sees" only the infinity of unconnected points, we are likely to bind ourselves together with traditions that compensate for our lack of connection by forming a type of social exoskeleton. We experience our unity only as members of a society: Christians, Freethinkers, Scientists, Buddhists, Philosophers, Methodists, Presbyterians, Rotary Club, Country Club, etc. As good members of these sociological structures that appear to organize us into wholes, we defend our teaching or convictions against all that challenges us. We believe that the defeat of our doctrine would herald the collapse of our unifying matrix and leave us once again isolated in the cosmos. Given this vision of cosmic isolation, it is virtually impossible for us to find the courage to brave our fears and to engage in the process that leads us to the love of knowledge. We settle for being defenders of our faith.

It was only in preparing for this paper that I took up again Tarthang Tulku's *Love of Knowledge,* a book that I had begun to read but had never completed. In Chapter Twelve, 'Transforming Time and Space,' I encountered the Star Vision. The Golden Flower, or the Multidimensional Cross of Christ, appears to be mirrored in the Star, an additional reminder for me that in the New Testament the spiritual presence of Christ in the receptive heart is described as the Morning Star rising (2 Peter 1:19). Venus, the goddess of raw instinctual impulse, is replaced by the agape, the Infinite Love that confers value upon all that it touches.

> Inner spark outer radiance
> Quality of being transformed
> Elements of Knowledge
> Alchemy of Time and Space (LOK, p. 88)

As a teacher, I am deeply concerned about methodology. Certainly, I affirm our need to cultivate the skills of inquiry, observation, analysis, and evaluation, but I am also convinced that we must allow our own personal stories to form a second focal point to guide and challenge our inquiry. I am committed to the goal expressed by Tarthang Tulku (LOK, p. 90):

> Apart from exploration through observation, analysis, and inquiry, we can cultivate a deeper

knowledge through respect for the possibilities of our own intelligence, and through a willingness to explore in all directions. This may require developing new symbols and a new language—metaphors and concepts that can do justice to all the circumstances and possibilities that inquiry discloses. But however we proceed knowledge is most likely to make itself known when we do not judge in advance what is worthwhile and what is not, or close off inquiry by relying on conventional views of what is advantageous to the self in satisfying its projects and needs. Being 'open-minded' in this way does not mistake openness for uncritical acceptance, but it does allow whatever appears to present itself fully for investigation by the inquiring mind.

## REFERENCES

Mascarŏ, J. (1978). *The Bhagavad Gita.* (NY: Penguin).

# ENGAGEMENTS

Because the TSK vision has to do with our lives, our experience, and our way of being, it does not submit readily to theoretical inquiry. It demands that we engage at every possible level the insights and presentations it offers and exhibits.

The most natural way to engage the vision is through a close and careful exploration of the exercises that appear in most of the books in the TSK series, or else through a careful reading and meditation on specific passages. The articles in this section all proceed along these lines. For the reader, this raises an important question: is it possible to reap the benefit of someone else's inquiry (i.e., the investigations that give rise to the articles in this section), or will real insight come only when the reader conducts the same sort of inquiry for him or herself. There may not be an either/or answer to this question, but it is certainly worth keeping in mind.

The first article, by Piet Hut, reflects on "Fields of Being," Exercise Six in *Dynamics of Time and Space*. This is the first of several articles that focus mainly on the knowledge dimension of the vision. The author reminds us that TSK exercises move easily toward enjoyment, and that the sheer joy of doing this kind of inquiry can itself be a gateway to deeper understanding.

He also suggests that the term "exercise" may be a too limiting. The investigations that TSK invites could just as well be seen as experiments and even (here the element of joy comes back into play) as entertainments.

The next article, by Tom Morse, looks at the very first exercise in DTS, called "Conducting the Vision." The author raises an interesting question: how can a single exercise (especially the *first* exercise in a book) embody all the themes of the TSK vision? And if it can, what is the purpose of doing all the other exercises? It is easy enough to craft theoretical answers to this question, but experientially the question is itself a kind of exercise, a form of inquiry that opens beyond itself.

The article that follows is taken from a talk I gave at a retreat held in Amsterdam in May 1991. It focuses on Exercise 5 in LOK. The exercise in the book centers on the experience of self as the owner of a situation, but the article takes a somewhat broader perspective.

Another LOK exercise ("Cloud of Knowing," Exercise 16) is the occasion for the next article. This article consists of the edited transcript of a practice session of members of a TSK study group. All the participants had recently taken part in a TSK retreat (referred to in the transcripts as the "Bello Retreat"). Their exchanges make clear that they are good friends, on a very familiar basis with one another. The transcript gives some sense of the special magic that can happen when individuals come together to practice TSK.

The group dynamics of TSK, and the power of practicing in a group, are well known to anyone who has taken part in a TSK workshop or retreat, so it is a special treat to be able to present the transcript of a group exploring together. It probably goes without saying that the session transcribed here should not be viewed as a 'model' for how a group does TSK work; other groups at other times will proceed very differently. The exercise that the group discusses here ("Cloud of Knowing," LOK Ex. 16) is especially appropriate for investigating how a group engages TSK, since it invites inquiry into multiple perspectives and non-linear complexity. The reader not already familiar with the exercise may not come away with more clarity as to how it can be practiced, but that is not the point. What the edited transcript shows is how discussion can be an exercise in itself, a way to deepen TSK inquiry.

The two articles that follow, both by Hal Gurish (also one of the participants in the Bello practice session transcribed in the previous article), present a different way of engaging the TSK vision: through a close reading of, and reflection on, specific terms in the text. The two phrases analyzed are "Body of Knowledge" and "focal setting." As is always the case with TSK, the analysis is suggestive rather than definitive. It offers possibilities for going more deeply into the text and the presentation of the TSK vision, and thus for going more deeply into the truth of being human.

The theme of 'focal setting' carries over into the next article, by Ronald Purser, which analyzes the exercises found in the Space section of the first TSK book. By surveying all the exercises and investigating their progression, the article is able to explore how various experiential themes emerge, develop, and take on new significance. My own article, which follows, complements Purser's nicely. Its focus is a single phrase from the Space section of TSK: "Space is projecting space into space!" I trace out one possible meaning of this phrase, which has struck a good many people over the years as incomprehensible, and go on to present four specific approaches to activating the insight it contains, based in turn on perception, embodiment, the world of the mind, and the ongoing commitment we make to the self.

The final article in this section offers an overview of the role that exercises play in the TSK vision. The author, Don Beere, systematically lays out the aspects of TSK exercises that set them apart from conventional experience, giving numerous examples to illustrate his points. His aim is to account, at least in a programmatic way, for the extraordinary effectiveness of TSK exercises, even for individuals without previous experience and without any kind of grounding in the TSK vision.

# Playing in the Fields of Being

Piet Hut

The Time, Space, and Knowledge vision invites us to consider everything that appears in terms of these three basic facets of Being: Space, Time, and Knowledge. And indeed, nearly all of the two thousand or so pages of the various books that have appeared in the TSK series are devoted to a description of the various relationships between appearance and Space, Time, and Knowledge. However, there are a few occasions where the lineage of these three facets themselves is traced back to Being.

One of the most striking of these passages occurs towards the end of the first TSK book. There, in the last chapter, we are invited in lyric prose to express Being through our own embodiment—to express the presence of Being through our own presence, in and as our own being. And on the last full page, we are invited to start right away, in a play-as-Being, without any need for preparation of any kind. No need to 'awaken' first; the keyword here is 'appreciation'. With appreciation leading to a play-as-Being, everything else can unfold in due time, and in some sense is unfolding already in the very act of a dedicated play-as-Being.

This chapter has always been one of my favorite parts of the TSK literature. No matter how often I have reread it over the years, I have always found pointers to new and unexpected angles, as well as connections with other forms of questioning I was engaged in at that particular time. Much as I have enjoyed and been inspired by each new book in the series, I suppose I have always felt a bit of nostalgia for that delightful crescendo at the end of the first book.

Against this backdrop, it was wonderful to be drawn right back into this chapter when I encountered Exercise Six in *Dynamics of Time and Space*. In this exercise, entitled "Fields of Being," we are indeed invited once again to sport in the fields of Being, playing as Being; at least, that is how it came across to me when I first read and did the exercise. Such richness, available right within experience—that constant companion of ours—available anyplace; anytime!

The recipe provided in the description of the exercise is simplicity itself. It can be summarized by one of its first sentences: "Let yourself enjoy the sense of relaxing the usual effort to conduct experience forward."

When I encounter a new TSK exercise, I often find it helpful to read the instructions a few times and then distill them into a single sentence. It could be a sentence taken straight out of the text, as in this case, or it could be one that comes to mind while tasting the instructions, rolling them over my tongue. I then start the exercise by paying careful attention to each single word in that sentence, to their interrelations and the associations that come up. Here is how I applied that approach, starting first with a brief note I made to myself when I did the exercise, followed by more detailed comments.

"What a delightful, cheerful, opening exercise! And so simple: first notice experience, then nudge yourself to the edge of experience, to the point where reality is conducted into being. Then notice the effort we tend to make in 'helping' reality conduct experience. Noticing how silly and superfluous such effort is, we can drop the effort, or at least relax it. And in doing so, we can naturally enjoy that state of relaxation."

The way I worked with the sentence "Let yourself enjoy the sense of relaxing the usual effort to conduct experience forward" was by starting at the end and working my way backwards. First I focused on simply noticing the presence of experience. It is always a surprise to see how easy TSK exercises are! What could be easier than simply paying a bit of extra attention to the presence of experience? Clearly, there is experience. We are asked to notice it. Since we are not asked to analyze it or try to transform it, nothing could be easier.

I then looked at the way experience changes in time, how it moves forward in time. Again, nothing to argue with, and a basic fact of life. Check experience: "Yes, still there." Check whether it moves: "Yes, still moving." But now comes the third step: "Check effort." Is there an effort involved in the movement of experience? This is a rather more subtle question.

Sometimes it seems as if everything just unfolds by itself, but most of the time it indeed does seem as if we are trying to help experience unfold: pushing it along, so to speak. Rather than simply being in experience as it flows, we somehow seem to feel a need to help move the world along. Even when we are just sitting in one place, doing nothing, a quick inspection of the muscles in our arms, neck, and shoulder often shows us that some form of tension is at work, slight, perhaps, but easily discernible if we care to notice.

As I observed this tension, it occurred to me that it was probably connected with our tendency to help experience along, our tendency to play conductor for the world as we experience it. This form of conducting experience forward does indeed take effort, considerable effort. I guess that we normally are not aware of all the effort we put into this process, simply because it is such a habitual and ongoing form of activity. It is like people living next to a waterfall and not hearing the sound of the water anymore.

Relaxing a tension once we have spotted it is not that hard. Spotting it is typically the bottleneck. So, having once noticed the considerable effort being directed towards directing experience, it is relatively easy to drop this effort, to relax it, at least to a significant degree. And once some relaxation is felt, what would be more natural than to enjoy it?

Still, we shortchange ourselves if we skip over the part of the instructions that asks us to enjoy. If we only try to "relax the usual effort to conduct experience forward" as an obligation, as an 'exercise' to do for ten minutes or so, we miss a lot. If we instead allow ourselves to enjoy the sense of relaxation that this exercise brings, it will more naturally carry over into daily life, after the time we set aside for the exercise has passed.

Partly for this reason, I have often felt that the term 'exercise' for TSK activities addresses only part of what is going on. Perhaps we could use alternative terms, such as 'experiment' and even 'entertainment'.

The term 'experiment' is apt, since we generally have no idea about the particular outcome of a TSK exercise. We are really experimenting with reality in a direct way (as the subtitle of the first book says: "a new vision of reality"). This activity is much more open-ended than the usual forms of exercise, such as exercising our memory or our leg muscles.

The term 'entertainment' may raise some eyebrows, since we associate this word with distraction and lack of concentration, and in general with relatively superficial forms of letting go. But a really good piece of entertainment does draw one's attention and ensures concentration. And 'letting go' is also a part of many a TSK exercise.

If we combine the notions of exercise, experiment, and entertainment, we can fuel our ongoing TSK investigations with a mixture of dedication, discipline, curiosity, and eagerness, as well as a sense of having fun. Actually, I have found TSK practice to be one of the most fun things I can do. Better put: practice merges with all other activities, and everything else becomes more fun. That is how I would sum it all up: Playing in the Fields of Being can be a lot of fun!

# Cultivating the Inner Landscape

Tom Morse

If you go closely into full awareness without preserving a realm of messages or concepts, you can—to use the language of DTS—"conduct openness." In openness, the root of not-knowing as a limit on knowledge proves hollow. Openness itself becomes the source of knowledge, and knowledge connects with a pure, awakened awareness that does not belong to the subjective realm.

"Conducting the Vision," the first exercise in DTS, can lead to a comprehensive exploration of awareness. I have often been surprised how quickly experience deepens through the practice of this exercise. It presents extraordinary possibilities, along with a sense of mystery. How can the entire TSK vision could be condensed and conducted within one exercise, let alone the first exercise in the book? What about the other exercises then; what is their relationship to the vision as a whole? Here I explore that question by trying to convey a sense of my experiences over many occasions of practicing this exercise.

* * *

'Awareness', like many commonly used concepts, is something we tend to assume we know or understand. "Conducting the Vision" asks us to look again. It begins with an invitation to explore awareness in its depth, to become aware of the awareness within 'each arising moment'. Already the question is there: "What is awareness, exactly?"

We are directed to look within each arising moment for the quality of awareness it carries. It is easy to overlook the implication that moments (in time) carry awareness. This perspective is fundamentally different from the common orientation that we as the observers are the ones who are aware. If moments carry awareness, awareness will always be present.

This is not our usual understanding. We tend to consider awareness a rather special mental operation. We say, "I was not aware of that!"meaning, perhaps, that we were preoccupied, that awareness was not present. The exercise challenges this assumption. It shifts our focus from looking for a thing or an object to something less definite. We attend to experience as movement, and this immediately makes our experience of space and time more open and dynamic. Now we may get some sense of the lived quality that is awareness. As we attend to the way that awareness transfers from one experience to the next, the quality of experience itself shifts. Clarity deepens. The boundaries and edges of experience begin to feel more open and spacious, relaxing one's relationship to the ongoing momentum of linear moments in time.

All of this—the initial questions, the experience of the moment, the continual attending to the moment, the appreciation of awareness being given or carried 'within' the moment, the transfer of awareness from one experience to the next, etc.—is awareness. With the recognition of awareness being transferred in each successive moment, our appreciation grows.

We can see this if we trace the practice from one moment to the next. At first we look for awareness in a particular moment, as though it were something to be found. Questions arise: "What is awareness? How will I recognize it being present or not present?" As we attend to this unknown awareness quality, with an openness that only the unknown can provide, the next moment arises. This moment appears qualitatively different from the previous moment. It seems to be more encompassing, since it includes much of the previous moment. The extension feels multidimensional. And it includes an aspect of the observer, the observing self.

* * *

Meditative disciplines often encourage being more aware. What is it to be more aware? Are there specific objects and attendings that would warrant the conclusion that we are now more aware? What is awareness like when we are not 'more aware'? Is it present at all? If so, what is its quality?

In the face of such questions, I find myself judging my experience of each arising moment. Something is lacking. I grow confused. What exactly is missing? But that question seems to lead nowhere. What matters is the tendency to make judgments of right or wrong, good or bad, progress or lack of progress. Such personal position-taking can be a challenge to awareness.

The exercise offers a different approach, a way to overcome the obstacle to awareness. It says: Attend to the quality of awareness, not to the content of what arises! As we attend to the way that awareness transfers from one experience to the next, any residue of confining judgment is relaxed .

We are looking at the quality of awareness. Often questions of quality are interpreted in terms of comparative judgments: good, fair, not so good. But 'quality' here is more practical and immediate. It points us toward experiences that we cannot find words for, but that may be valuable in appreciating how experience unfolds and infolds. 'Quality' can become a means for becoming familiar with awareness itself.

## Inclusive Awareness

Awareness is not static nor contained, so looking for the presence of awareness can be difficult. One may get a glimpse, but the mind, set on locating, identifying and pinning down, finds it difficult to maintain this.

The focus on how awareness transfers from one experience to the next helps. It lets us appreciate awareness in a more open manner, not reduced to an object. A more subtle and broadly based knowing emerges, one that includes, for example, the 'field' around objects, thoughts and perceptions. Awareness shifts from the content of what appears to the background and edges of appearances, from "What is it?" to "what is it like?" The rigid, fixed nature of appearance begins to relax or open, and awareness moves beyond the confines of content into the 'field' of awareness.

Field awareness can become somewhat hazy and dreamlike. A sense of ease accompanies it, but this is different—separate from—the fullness and immediacy of experience. The directions for the exercise address this by asking us to look 'within' awareness to discover the 'knowingness' it offers. As we 'fine tune' in this way, awareness engages the particulars within appearance, bringing alertness. Worlds of experience begin to open within what we take as 'the ordinary'. Insight unfolds, going beyond content. The structure and dynamics of ordinary experience begin to reveal themselves, allowing space to open further. The observer, the active knower, begins to experience a sense of freedom and ease, relaxing the 'need' to know in favor of appreciating a more pervasive knowingness.

In this experience, knowledge opens in surprising ways. It begins to seem as if knowledge is freely available within whatever appears . . . or does not appear. Eventually, this newly expanded awareness begins to engage what we might ordinarily refer to as 'no awareness'. We recognize awareness where it appears to be absent, and this awareness of 'no awareness' is then continued into the next moment, and the next, and so forth. As awareness continues to expand into areas where it had previously seemed to be absent, I experience a sweet abiding that does not immediately draw away from pure experience back into the familiar. As appearance and awareness become the same, nothing remains outside. 'Letting go' is no issue and requires no effort. The knowing 'within' allows an unlimited display of possibility.

## Full Awareness

The exercise continues: "Be careful not to tell yourself what kind of experience to have or dictate its nature. You can swim in awareness; you can refine it as you embody it. The process develops on its own, without special effort on your part."

Swimming in awareness deepens the quality of awareness. Whatever arises—my hand before me, the pen writing on paper, my perspective on what is occurring, the observer as the subject pole of experience—is appreciated as awareness. Space opens to accommodate more fully the multidimensionality of experience. Experience ranges from the ordinary linear and sequential to a more dream-like phantasmagorical quality. They 'are', yet they 'are not'. I 'am', yet I 'am not'. 'I' and 'they' can be appreciated as awareness itself.

The commentary to the exercise presents an extraordinary statement: "The more you practice the exercise, the more you will recognize that the aim is not to find new knowledge or to succeed in doing what is asked." What are we doing if not trying to succeed? The commentary continues: "This way of structuring the exercise will only confirm the view that the knowledge we ordinarily operate with is incomplete." There is something more important, it seems, than success. The knowledge that relies upon progression reflects a seeing that lacks fullness. It is a restricted knowledge.

The text continues: "Here the challenge goes deeper. If you go closely into full awareness without preserving a realm of messages or concepts, you can conduct openness." What is full awareness? When is awareness not full? Pursuing these questions is tricky, since whatever appears or does not appear can be appreciated, both through and as awareness. Yet, the quality of awareness does appear to vary, to be different at different times. The pursuit of these questions becomes problematic. It seems to go nowhere. But the text offers wonderfully incisive guidance: "In openness, the root of not-knowing as a limit on knowledge proves hollow. Openness itself becomes the source of knowledge, and knowledge connects with a pure, awakened awareness that does not belong to the subjective realm."

Trying to follow a particular procedure, I find myself stuck. What is it about this particular procedure that is limiting; that is not open? The point of conflict centers around the difference between the appearance of two distinct experiences. I am applying the focus of ordinary knowledge, which looks at the content of appearance rather than its quality. I find myself concluding that something is lacking in one experience as compared to the other.

This is what the exercise warns about: attempting to preserve a realm of messages or concepts. The surface appearance or content of the two experiences has been taken to represent the whole of the situation. The 'something is missing' relationship has been given special status. But now I have seen this tendency. Appreciated as awareness, it becomes incorporated into the 'swimming'.

Tarthang Tulku (DTS, p. 246) refers to this divisive and alienating tendency in relation to the 'role of the seeker after knowledge'. The seeker activates a "self-reinforcing drive." The passage continues:

> Intent on perpetuating the surface of 'appearance as substance', mind reacts to any challenge by affirming a 'source-from' understanding. It may trace what appears into the subconscious or unconscious mind; it may pursue one of the countless connections and associations that thoughts evoke. Alternatively, it may lapse into blankness, disclaiming any knowledge of how to research or investigate inwardly.
>
> Once we take on this way of being, we get more and more involved, like someone who begins to hyperventilate. The pathways of knowledge are saturated with hunger and desire and fantasy. Thoughts and senses all trend in a certain direction. We can only look from and to; we can only wait and wish, and then wait some more. In the end there is frustration or disillusionment, or else more anxiety.

Full awareness appreciates this tendency to hold to an apparent realm of messages or concepts as itself being an expression of knowledge. This appreciation is the 'full swimming' of the exercise. Whatever arises is simultaneously the means and the medium of experience. The self, being within the realm of preserved concepts, holds no special status. The outcome may be confusion and frustration, strains and dramas. But full awareness invites all presentations to join this drama, to offer their wholeness.

Would it be possible to 'live within' the practice of Conducting the Vision? If so, I have not found the way. Invariably experience seems to return to the ordinary, where problems need to be solved, where something is missing, where solutions are not 'here' but lie somewhere else. Seeking answers to our questions, we are drawn back into the ordinary.

## No Exceptions

Within ordinary experience, we seem to enter gray areas where awareness takes a rest. We accept the reality of the parameters we place around and within the given experience. The unacknowledged message is, "This is the way it is . . . . This is the whole truth!"

Awareness can bring light to this 'gray rest-area'. It asks, "Who says so?" As with all TSK questioning, the act of inquiry makes a more comprehensive and revealing awareness available. It brings fresh perspectives to the question, "What is real?"

Chapter 28 in DTS, "Silent Appearing," poses the 'problem' of ordinary experience. It begins (p. 277):

> Our ordinary focus on the content of experience as it arises in time is something like focusing on the air bubbles that form within a rushing stream of water. It does not occur to us to ask how the bubbles arise, or to imagine that we could trace this arising. The bubble's content encompasses our existence, our experience, and our thoughts, and we have no interest in penetrating beneath these levels.

How is it that we lack interest in knowing directly the nature of our experience? How can we activate interest where it does not exist? Once again, focusing on the 'quality' of awareness rather than its content offers a doorway into the exploration of this dilemma. We are looking into the 'problem' of having no interest. But instead of looking at the content of that problem and trying to work out a solution (DTS, p. 277–78),

> [W]e can investigate the quality of the observing mind. Sensitive to the active knowing that guides our efforts, we can train the body of mind. Once we understand how we conduct patterns in the body of space and time, we gain more direct access to the Body of Knowledge. Newly knowledgeable, we can deal with the emerging problems in our lives at the very outset, relying on our own understanding rather than knowledge possessed by others. We can open to the presence of awareness and make the choice to embody our circumstances in new ways.

How can we not be interested in this knowledge? What are we interested in instead? Looking at our experience, we see that we are continually influenced by the content of what appears to our mind. This, along with the compelling nature of the past-present-future momentum of our experience, leaves little room for assessing what we might otherwise find truly valuable and deeply interesting.

Are we willing to acknowledge that we really don't know how events show up in our lives? Are we willing to put immediate application and problem solving in the background and explore the nature of the mind itself? The ordinary mind lacks the motivation or interest to look at such possibilities. But attending to the "quality of the observing mind" embraces the unknown and opens space. This openness invites possibility, while awareness touches new ground.

Tarthang Tulku writes (DTS, p. 278):

> Just as water can appear as clouds or ocean or ice or in a rushing mountain stream, so awareness can take form in conscious mind, the unconscious, thoughts, emotions, perceptions, joy, insight, or virtue. Awareness sponsors our assumptions and shapes our experience. It is active not only at the outset, not only in naming and defining, but in each arising moment.

Once awakened, awareness, can illuminate every part of our experience. Active in the background and the foreground, in tension and relaxation, in the self-openness of seeing and the darkness of not-knowing, awareness discloses the operations of mind and the ongoing interactions of the senses with their objects. It protects us from stumbling over obstacles that our own way of minding creates.

Usually we rely on rational activity and the reasoning mind to get us through the day. But if we begin to explore the unknown within our experience (see "Choosing the Unknown," DTS Ex. 23), we find it wherever we look and within whatever appears. Even certainty is permeated with the unknown, as we see when we inquire into how we know our certainty. If we function as well as we do, it may be because we simply ignore the obstacles that continually arise within the framework of our cognition. In the midst of contradiction, we hold firmly to the reasoning mind.

Awareness deals with these patterns by appreciating the space within our rational activity. In this simple way, the seemingly fixed and rigid quality of appearance, along with the logic that supports it, become transformed. They open beyond the confines of the linear and conceptual. Attuned to the quality and the field of what appears, awareness invites a seeing that goes beyond the dark not-knowing of the ordinary.

Exploration and inquiry continue through the gift of the TSK vision. Conducting the vision unites us with experience itself, revealing an endless unfolding of our human embodiment. As rich and subtle nuances continue to appear, one begins to appreciate what TSK refers to as the Body of Knowledge—the endless display of life's treasures.

# No Inside, No Outside

Jack Petranker

When knowledge is directly available in experience,experience takes on a quality of richness, of completion. Everything just fits. This kind of experience has a jewel-like quality. What some of you said about your experience yesterday morning had that flavor, which is good, because once you taste that flavor, you don't forget it. Or rather, you do forget it, but when you come across it again, you remember; you say, "Oh, yes. Right."

In this rightness, there is a basis for trusting experience. When that trust is there, you may find yourself doing something without being sure why, but gradually it becomes clear what is going on. You don't have to figure it out. You might describe it as being in harmony with the situation as a whole.

None of this depends on picking and choosing what is good or helpful. So often we think to ourselves that our experience is partly good and partly bad. There is a way of understanding in which it makes sense to talk this way, a very important way that lets you make distinctions and discriminations about what has value and what does not, what will produce good results and what is harmful. But there is another way of understanding where we don't need to pick and choose. On that level what seems most valuable is to be clear, or to be fully present to the range of experience, without turning it into objects or even patterns or tendencies.

Let us be more specific. Suppose you are having an interaction with someone, and in the course of that interaction you see that you have a need to be liked. So as you are speaking with that other person, you are also noticing, "I want this person to like me." That knowledge is a part of the completeness of the situation. It helps bring the situation to perfection.

The more complete your knowledge of what is going on in a situation, the more effective that knowledge can be. For instance, in the example I just gave, I still referred everything back to myself. That is also a limitation on knowledge, a story we tell that sets limits and narrows down the field. Awareness of that limitation is another aspect of the perfection of knowledge. In *Love of Knowledge,* Rinpoche speaks of taking the knowledge available in the moment together with all the limits on that knowledge, all the blind spots, because they are all knowledge too. This is a "1+1" that always adds up to infinity. Not some special infinity that is glorious or beyond experience. Just everyday infinity: the whole of the situation, without distinctions—or rather, with distinctions, which are now just a part of the whole.

Knowledge plus the limits on knowledge doesn't exclude anything. So if you can fully take in, fully 'embody' the limits of your knowledge, there is no additional knowledge required. This may sound like a trick with words, but if you look at your own experience, it begins to make sense.

One exercise that has been helpful to me in this regard is "Disowning Tension," Exercise 5 in *Love of Knowledge*. We won't do it today, though we will do something similar, but I want to describe it, in my own words. Perhaps I am really describing a variation.

I think of this practice as "no inside/no outside." To see what I mean by this, let's look at a typical way we might have of investigating experience. We might say to ourselves, "Now I want to be fully present to my situation." Usually we understand this to mean that we are going to cut out the mental chatter and make sure our senses are very open and receptive, and that in being with other people we will be very receptive as well. If we do happen to start thinking of something else, we might say to ourselves, "There I go again! I'm back in my old patterns, the ones that keep me from being present in the situation."

This way of understanding what is going on depends on putting into play a particular model of experience. In this model, we acknowledge a tendency to get caught up in what's going on inside us, inside our experience. This tendency keeps us from being fully present.

It may be the thoughts we are having/ but often there is an emotional tone. For instance, you have a conversation that upsets you; later you are walking in the sunshine, enjoying the day, when you remember the conversation. For a minute or so you lose track of everything else, and the good feeling of the day disappears. Does that sound familiar?

WIth reference to this example, the practice of "no inside/no outside" says that there is no difference between those two circumstances: between being present in your outer environment and being present in your emotions. Or rather, there is a difference, because being *caught* in your emotions is not being *present* in your emotions. If you were present, the emotion might not stay an emotion. But that just confirms the basic point of there being no real difference.

The world of the senses and the world of thoughts and emotions are both environments. The second-level stage of space and time and knowledge equally allows each of these environments. One way to think about this is that the emotions and thoughts are 'out there', just like the world of sense experience. The other way is to think that the physical world that we know through the senses is 'in here', since it shows up in experience in the form of sense impressions flowing in, just as emotions are mental impressions flowing in.

That is the practice of "no inside/no outside." The two are equivalent. What is outside (the 'real world') is inside, and what is inside (my emotions, thoughts, moods, etc.) is outside. Since that is so, you don't have to fight against what is going on in your mind, because your inner experience is not a distraction from present experience; it *is* present experience.

This is tremendously liberating and relaxing, because it means you don't have to reject thoughts and emotions. At the same time, you don't have to get so caught up in what is going on in the mind, because it occupies no special position. It is no big deal. It is just more activity, more 'things' presenting themselves in the field of second-level space. That is what I mean by "no inside/no outside." The categories, the sharp sense of division, just do not apply.

Now Exercise 5 in *Love of Knowledge* applies this in a more specific way. It makes the suggestion that you can take the self as part of experience, 'out there' along with everything else. And that is also true. The self has no special, privileged position. There is no space that the self alone occupies. It arises together with everything else. It is not special, and it is also not 'wrong'. The self is not the enemy. It is just more activity, just more things showing up in the flow of time and the range of space, known by knowledge in terms of certain identities.

Before we were talking about plans, and the same analysis applies to them. A plan is just as much a part of the environment as the sun shining through the windows. That is pretty remarkable. It tells us something about second-level space, and it tells us something about how to be.

We will do some practices this morning related to all this, so we can see how the practice unfolds.

# Cloud of Knowing

## Bello Retreat Group

Tom: Which exercise shall we do?

Hal: I was thinking "Cloud of Knowing" [LOK Ex. 16] would be appropriate.

*[editors note: Exercise 16 (minus the illustrations, which show various lines at right angles), reads as follows: "Ordinarily, we look at objects, forms, shapes, and events from one perspective at a time: a simple linear relationship that can be understood either temporally or spatially. Viewed more closely, observation does not follow this simple, linear path. As we observe one thing, other thoughts, images, or judgments may come to mind. Again, the object we are observing may reveal new aspects or dimensions, or we may see it from different perspectives. Observe perceptual situations with sensitivity toward this non-linear complexity."]*

[Forty-five minutes practice. Three minutes before the end of the practice, Hal quietly suggests that we gradually come out of the exercise and notice what is there for us as we come out.]

Hal: I certainly could feel a change as I was coming out of this exercise. It's hard to describe. The coming out was literally a coming out of something, out of some kind of 'in', which was maybe death or . . . I can't say. It's still present a little bit.

Doug: When you said that about noticing how we come out of the exercise, it became a visual thing, as though I was coming into my body—into flesh and pinkness. Into my eyes, as though maybe I had been . . .

Hal: Maybe outside of your body?

Jean: To me the 'in' seems so spacious, in a way that feels like light.

Hal: You mean like inside turned out?

Jean: It feels like I'm going 'in', but it feels also like I am going 'out' at the same time. When I come out into this room, it feels like it is more 'in', because it is more limited by what I see.

Hal: By the boundaries?

Jean: By the boundaries of that flower, for instance. When I'm 'in' there's much more to that flower arrangement than when I come 'out', which is in this room.

Hal: Something to do with the quality of seeing?

Jean: Yes. In and out, like the Doctor Seuss book. It's such a mix-up.

[pause]

Lucy: I was surprised how fast everything changed!

Hal: When you were coming out?

Lucy: No, just when I was there, doing the exercise. How one thought or one image would so quickly change into something else. I think I must have gone all around the world. Rapid change, but at the same time I was slowing down. The images were like glittering light or quicksilver, and at the same time another aspect of me was slowing down . . . or maybe opening to a larger space. It was a movement of slow and fast at the same time.

Doug: Slow and fast at the same time: does that mean flipping over—slow/fast, slow/fast?

Lucy: No.

Jean: It's not really that you are slowing down, but the images are speeding up. Do you get that? [laughter]

Lucy: That's a good stretch!

Tom: A right-angle stretch.

Jean: I know what you're saying, Doug. The first time I ever really deeply felt "expanding and condensing" was at a retreat. I'll never forget that, that duality, expansion inside of condensing.

Marion: Saying it messes it up! Because it happens at the same time. I've had that experience too, but when you say it, it's a mess. It's spread out and linear, and just that way. And it's not that way.

Jean: It's at the same moment. Inside . . .outside . . . .

Marion: It's just there. It comes right into your chest.

Lucy: Going back to "coming out of the exercise . . ." We went into the exercise and then there was some experience, and then we came out of that experience. But maybe we didn't come out of it. Maybe we're just in another extension of it. I don't feel like I came back to anything; I just went into something different.

Hal: You're in a different space now than before you did the exercise.

Lucy: Yes. I didn't come back.

[pause]

Tom: This exercise confused me in a way. Sometimes I think I know what these exercises are about, but this time everything was in reverse. First of all, it took me a long time to get 'in', whatever that means, and then it seemed that the cloud had more to do with my just staying in the ordinary. I was kind of hanging out, like a whole cloud, staying in the ordinary, looking around. And I see something . . . "What's this going on? Oh, change." Nothing was really that different. It felt very cloud-like. And then I just started to follow my breath. I thought the quality would be more expansive, or more on the verge of something opening. But the cloud was just like a cloudy day, totally encompassed in the ordinary. Very uncomfortable, the ordinary! I guess because I thought I was doing it wrong. So it was very uncomfortable staying there.

Then it moved into something different, and that seemed more stable. More peaceful, more expansive, more what I expected. And then the coming out of the exercise: it seemed I didn't have to go anywhere, because I never left. It seemed I was back on the cloud again. It was just totally reversed. . . so totally ordinary!

Lucy: What was 'ordinary' like?

Tom: Well, it was really abrasive.

Hal: The process of the exercise? It was abrasive because it was ordinary?

Tom: Well, I don't know [laughs]. It seemed so extraordinary that the ordinary was like a cloud! Normally, I would think that the ordinary would be very clear. That's as clear as I can get with words. The ordinary is where I live my life! Where this is 'this' and that's 'that'. But it didn't seem that way; it seemed totally different! That's why it was abrasive. It was only when I could go 'beneath' that that it got more comfortable. Whereas normally I would think that my comfort would be in the ordinary. Everything just got turned around, topsy-turvy. I don't know what to make of it. I think it was abrasive in part because there didn't seem to be any entryway to a broader awareness or knowing, like all the arrows in the illustration to the exercise. There didn't seem to be any possibility of things being informed in different ways, or penetrating. It was so condensed.

Jean: What strikes me is how you had a plan and it just totally . . .

Hal: Got sandpapered away!

Tom: Well, I don't know what my plan was but obviously, there was something, becausc it was juxtaposed to what was happening. What was happening wasn't right—in fact it was totally wrong! I was doing something wrong and I didn't know what 'right' was, but this wasn't it!

Jean: Maybe that was a right angle, right there.

Tom: It certainly was! It's interesting to try to speak to that right-angle experience, and also the resistance of speaking to it.

Jean: Grating.

Tom: Yes. I didn't really want to talk. Part of me wanted to bring it up to see what came out of it. But there's also the part of me that still thinks I know what I am going to say before I say it . . . and that's the way it should be.

Doug: Like it's not authentic unless you think it first!

[laughter]

Doug: Not verifiable.

Tom: Part of me thinks that I do operate that way.

Doug: Does it seem that you should think something through before you say it? Is that an agreement that we have with each other?

Tom: I don't know if that ever really happens, but my mind still thinks that way.

Doug: My sense is that we have a kind of social agreement that you should think something through before you speak your mind. It's like being a person of your word.

Jean: But you know, sometimes when I feel very connected to deep space I don't feel like I think it through at all.

Tom: Well, I thought that was what our laughter had to do with . . . the realization. . . . For me, I realize that I don't think I ever, even now, think of it as I am saying it, but the saying is coming and it's coherent. That's what I thought the laughter was. We can't think fast enough to be ahead of our speech.

Marion: We assume that we do. But I want to ask: You said you went 'beneath', not 'opposite', but beneath.

Tom: Oh, that's interesting! It wasn't opposite. The exercise didn't seem to work, but changing it wasn't like switching to the opposite. What I had to do was more like [the Kum Nye exercise] Swimming in Space. It was a kind of sinking down into this abrasiveness. And then it seemed to be more calm or steady, more peaceful, not abrasive.

Lucy: A very strong, a very evocative word!

Tom: What's that?

Lucy: Abrasive, like sandpaper, rubbing, it has a real texture to it!

Tom: It wasn't a common feeling for me.

Lucy: So it was a surprise?

Tom: It was a surprise. A lot of this exercise was a surprise for me. I don't often have a sense of things being totally wrong. Really wrong! That's abrasive. If it was like sandpaper, it was coarse, not fine.

[Pause]

Lucy: Do you feel that sensation right now?

Tom: Yes, even in the recall I seem to have brought it back.

Hal: What I feel is that you are looking at me so intensely.

Lucy: Yes. There is something in the texture of space. Maybe other people feel something textural or something that connects with the sensation that Tom's talking about.

Jean: Hal picking up on the eyes is part of the connection. It's in the intensity. That's what I appreciate about the group.

Lucy: It's in the field; there is something that actually enters the field.

Tom: I'm more aware of this quality now that we're talking about it.

Jean: In your body?

Tom: Yes, but not as strong.

Lucy: There's something strong, though, something I can relate to.

Hal: Something to do with the eyes . . . .

Lucy: Maybe the eyes have something to do with the looking . . . .

Hal: Jean mentioned that her way of looking changed when she came out.

Lucy: It's a looking softer, more diffused.

Tom: What comes to mind is the right angle. That's abrasive sometimes. The whole thing looks that way. My whole description . . . .

Jean: You did explain that in terms of "Cloud of Knowing," with all the right angles and left angles. You got whacked with that right away. That's an unknown!

Tom: Well certainly it is, but I thought the exercise was within what I did. I just couldn't understand it or make sense of it. That probably explains the abrasiveness too. It's interesting to have something be like a right angle, to be there in a way that is so predominant. Usually, I have to conjure up these right angles.

Jean: How can you conjure up a right angle? I can't even relate to that.

Hal: What I did in the beginning was almost look to the right. Like for that bird in the picture on the wall to come in.

Jean: With a message?

Hal: Yes.

Jean: So you 'do' the right angle, mentally. I never have to do right angles—they always surprise me.

Tom: Somehow I get a linear sense of flow of the way things kind of enter me, and when I think of 'right angle' I just let it come from elsewhere.

Jean: Well, it did come from elsewhere

Tom: That's what I mean. But that's what I do.

Jean: But this time you didn't. This time it came from elsewhere.

Tom: I didn't do anything to this one.

Jean: And you didn't recognize it.

Tom: Not the right angle, no.

Lucy: And you followed the right angle.

Tom: It doesn't seem that way from some perspectives. If we hadn't had this discussion and brought it up and contained it somewhere—held it—I wouldn't ever have had the sense that it was a function of the exercise at all.

Jean: Well, I accept it as closer to the truth, because it caught you off guard.

Tom: You know, my mind wants to have something to know about it, something I can own. But just that I have the experience . . . That's so rich in itself. Because usually I discount that. It's refreshing. It makes me wonder how often that is present in my life

Jean: Without your appreciating it?

Tom: Yes. At right angles to my experience.

Lucy: If you look at the diagrams in the "Cloud of Knowing" exercise with all the little arrows—where are the clouds? Why aren't there little puffy things coming out of the ends of all those arrows? This is a very linear description of something that doesn't have any lines in it at all. And it doesn't have any little points on the ends either.

Tom: That was my reaction when I looked at the diagram.

Lucy: I've heard the exercise a billion times, but I've never seen this little diagram. So it was a really different experience. The arrows are symbols pointing at clouds!

Jean: That's right. How can you draw lines in a cloud of knowing?

Lucy: Maybe it's a linear description from a mind that conceives of clouds as arrows.

Marion: The last line [in the exercise], "Observe perceptual situations for sensitivity toward this non-linear complexity." There's a lot in that sentence!

Tom: And the non-linear is pointed out to you with all these straight lines.

Marion: That's what I'm staring at. That's what we operate with—that's how we move in the world.

Tom: "Linear complexity" Could that have to do with many things happening at the same time? Which linearly can't happen.

Marion: But in a cloud there's never just one thing happening.

Tom: But he [Tarthang Tulku] shows all these different things happening with arrows. And of course my mind is trying to understand it. Trying to understand a very linear complexity.

Hal: With your ordinary mind; with your ordinary way of thinking.

Tom: Yes! And there's the abrasive quality!

Marion: And you go beneath it to try to see it in a non-pointy, non-scratchy way.

Tom: To go into the movement somehow. It's interesting taking that thing apart, clouds and arrows, how that works. When I see the diagrams I just want to turn my head [laughs].

Lucy: We can reinvent the diagram as a non-linear description of the cloud of knowing, with puffs of clouds going off all sorts of ways.

Tom: The way you see it, all these points have puffs, like balloons in a Charlie Brown comic strip. I saw the whole thing coming out of a puff.

Hal: The whole drawing.

Tom: Yes, the whole drawing.

# Body of Knowledge

Hal Gurish

How does the Body of Knowledge (which plays an important role in KTS) differ from Great Knowledge or from 'knowingness'? A starting point is the Body of Knowledge as it is introduced in the first book (TSK, pp. 254, 287, modified):

> Everything is known, and known as the unity—the intimate union—of Time and Space. . . . Everyone participates equally in this Body of Knowledge. Everyone and everything is the total and vigorous embodiment of Space and Time's intimacy. We can call this total communion the Body of Knowledge.

I have found it helpful to think of this communion as the coherence of knowledge, an all-pervasive communication that provides order, without restricting that order to a single form. So the Body of Knowledge offers a multidimensional ordering coherence.

This suggests that we should be careful not to reify the Body of Knowledge. It is not a body, but rather a working together. It is interactive communication, without there being 'things' that interact. Through this interactive dynamism manifestation can take place and things appear. And since this is so, all things share in the Body of Knowledge, much as any point in a hologram presents the whole of what the hologram records.

We speak of the physical body as a single entity because all its parts work together as a living organism. In the same way, the Body of Knowledge is a living organism. How can we touch this aliveness? Tarthang Tulku suggests that one way in is through the senses (KTS pp. 441–42, 447):

> We can touch the Body of Knowledge by focusing on our embodiment directly. With our senses attuned to the world around us, we grow aware of a more encompassing body of knowledge. Expanding awareness aligns with the subtle energies of inner experience, making knowledge available not only through the activity of the senses and mind, but throughout the realm of physical space and linear time. Rigid dichotomies begin to dissolve and loosen in favor of a more open allowing and a more active presentation.
>
> . . . We can participate, we can own, we can use, we can play—for all doing and all being is the 'aliveness' of the Body of Knowledge.

These passages suggest a way of making the transition from individual embodiment to what might be called the total embodiment of the Body of Knowledge. The senses are communicative and responsive. Although they particularize, they are also responsive. This particularizing responsiveness can expand, becoming non-particular and all-encompassing.

It might seem that what is non-particular cannot also be coherent. But if the Body of Knowledge is concrete in a way that allows for multidimensional orders, there is no conflict. Perhaps a quantum leap occurs when the particularizing tendencies of embodiment expand toward infinity. The following quote (KTS, pp. 446–47) points in this direction:

> [C]onfinement itself is only the characteristic indication that lower levels are in operation; from a 'third level' perspective, confinement has no reality as such. When interpretations of lower-level knowledge are understood as presentations of the Body of Knowledge, 'knowingness' remains fully available even when such interpretations are in effect.

The TSK vision suggests that we can concretely touch the all-encompassing Body of Knowledge. My own experience hints that this is possible, perhaps in those moments when everything 'makes sense'.

Here a note of caution is appropriate. In exploring this possibility, it is natural to start with a 'doing' directed toward a goal. However, this orientation makes the self as inquirer into the subject and reduces the Body of Knowledge to an object. It particularizes in a way that closes down potential.

A better approach is to allow for participating in the Body of Knowledge. In that case, there will be no possession and no attainment. Only in retrospect could we even say, "Yes, at that time I was touching the Body of Knowledge." And perhaps in the end, this holding will have to be abandoned as well.

# Focal Setting

Hal Gurish

The term "focal setting" in TSK refers to a perspective, or frame of reference, or level of analysis. It is connected to the notion that the kind of knowing we employ profoundly influences the reality we perceive and move in. By focusing in a certain way, we construct our world. A narrow focal setting means a small perspective, while a wider focal setting refers to a more open perspective.

An example of a narrow focal setting is viewing the human body as a solid, separate entity, completely confined within the skin and isolated from its surroundings. This 'view' is not just activated in our way of thinking about ourselves, but in our lived experience of our situation.

With a wider focal setting, we might see the body as an enormous complex of molecules and atoms. Here the body has lost its intractable solidity; it has become mostly space. An even wider focal setting would allow an understanding that individual events—atoms, cells, human beings, planets, solar systems—are not separable but are interconnected; that the boundaries between events are a product of the particular focal setting in operation.

A more narrow focal setting normally goes along with our everyday view, the one we take for granted. It is supported by a background of belief, and it could even be seen as the congealing of belief into particular cognitions, perceptions, theories, or ideologies that shape experience. But we can learn to question these habitual perceptions. For instance, when we look at a table, are we certain that this table is the same table from day to day or moment to moment? Or are we seeing a constant table because that is what believe is there? Solid, stable objects express a belief. They summarize momentary experiencings in the expected way.

Understanding how focal settings work can alter how we view ourselves and our lived situation. Problem areas may open up when we see that that defining characteristics that 'spell out' the existence of a problem are the outcome of a narrow focal setting. With a wider focal setting, our problem might be an opportunity to re-examine what we have taken for granted. Pain and illness, for example, might be a signal to review the way we have been living. Confusion could be a challenge to outworn beliefs. Even an accident might serve to wake us from a dull, stony complacency.

The everyday world and its problems are not excluded by a wider focal setting. A wider perspective is not 'spaced-out'; rather, it includes ordinary events, seeing through them without denying them. A truly open focal setting, one that penetrates the implicit assumption that the world is objective and independent, can reveal the illusory nature of our everyday reality and still appreciate its value.

Opening the focal setting requires questioning the assumptions that structure and dim immediate experience. For instance, we view experience in time and the changing nature of events in terms of a focal setting that identifies the self. Could a wider focal setting let go of this belief in 'my identity', seeing an ongoing process, re-created moment by moment? Could the memories that establish my identity be seen as summary solidifications of the dynamic unfolding of time?

A focus on time's dynamic offers a valuable opportunity to investigate focal settings. If we can learn to notice both the more subtle and the larger sweeps of time, the process of change itself can be seen in a new light. The usual focal setting leads us to wish for certain changes and fear others, with the result that we blind ourselves to the change that is ever present. The magnitude of that change gives access to the power of time. Like a wrestler taking a solid stance, or a skillful boatman shifting his weight this way and that in accord with the flow of the river, we can experiment with different, more open focal settings to change completely our experience of the world we inhabit.

# Questioning Space Through Experiential Practice

Ronald E. Purser

The first TSK book is divided into three main sections: Space, Time and Knowledge. Each consists of analytical investigations and experiential exercises designed to unfreeze the rigid presuppositions that ordinarily influence our embodiment of these three facets of experience. I believe it is no accident the starting point is a focus on Space, for a new appreciation and understanding of Space seems central to the vision. Reflecting on my own early encounter with TSK in 1982 during a ten-month program offered at the Nyingma Institute, I believe the Space exercises had (and still have) a transformative effect on my mental outlook, embodiment, and emotional disposition.

In the Space section of TSK, a progressive series of exercises complemented with commentaries and text serve as a foundation for achieving a shift in "focal setting." The first series, referred to in the text as the "Giant Body" exercises, employ imagination, creative visualization, and observation. The second series involve attending to familiar aspects of experience (such as mind, body, feelings, and thoughts) by seeing them in their Space aspect, along with questioning the location and source of thoughts and other forms of mental experience. The last set of exercises in the Space series deepen and integrate previous insights through challenging subtle habitual tendencies and by working with more intensive gazing at an open and clear blue sky for extended periods of time in retreat (Exercise 15, "A Mountain Retreat").

Tarthang Tulku tells us in TSK that Space is very much the nature of our embodiment. By learning how to appreciate Space more, we can enjoy a more open and spacious sense of being. However, our usual way of relating is ". . . often very rigid, which restricts our potential as embodied beings" (p. 25). It is important to point out that embodiment includes our view of reality, that is, the way we know. In fact, ". . . some kind of knowing is a primary fact of embodiment" (p.24). The implication here is that a more primordial way of knowing, grounded in a more space-like embodiment, is available. This primordial knowledge is not predicated or limited by our usual habitual structures of knowing, which have rigidified into a split between self and world. Our habitual focal setting (which we embody and live out), presupposes a fundamental duality between subject and object. The TSK exercises and inquiry are designed to 'thaw out' or unfreeze this rigid, dualistic focal setting, allowing ". . . a greater measure of liveliness, clarity and intimacy" (p.25).

## The Giant Body Exercises

The "Giant Body" exercises are designed to open up our perception of the body, which is usually felt to be dense, rigid and solid. The exercises begin by having us imagine a Giant Body, seemingly projected above us 'out there'. However, I have found it very helpful to imagine the Giant Body as my own, and not as some separate image projected at a distance. The instructions to the exercises also ask us to develop a measure of accuracy and detail with regard to viewing the various structures, sub-structures and micro-structures.

These exercises do require some discipline and training in concentration. CD-ROMs are now commercially available that provide a 3-D illustration of various components of the human body, a helpful support. It would be interesting to know whether TSK practitioners who already have detailed knowledge of human physiology have found their practice of the Giant Body exercises 'easier' or more vivid compared to those who lack such familiarity.

A number of people attempting to conduct the Giant Body exercises have expressed difficulty with visualization, and in developing precision and stability with regards to details of the body's systems. My advice to them is that they should not think too narrowly or rigidly about what visualization is supposed to entail. It might be more fruitful to conduct the exercise in a more playful fashion. For example, what if we could experiment with allowing our imagination to imagine the body's systems in full detail without any mental effort? Or perhaps we can use the power of emotion or feeling to build up the image. We might even imagine that we are sitting in a large movie theatre about to watch the best special effects that Hollywood has to offer. We should also not be discouraged if we cannot see a clear visual image at first; feeling the 'feel of the image' may be enough and provide entry into a different way of knowing.

I have always had a fondness for Exercises 4 and 5 in the Giant Body series, "Just Interactions and Shining Outlines," and "Released to Space." These two exercises help to counteract our ordinary perception that things (in this case the body) are permanent, solid, and self-existing entities, inherently and independently existing from their own side. Instead, we develop a new perceptual capacity to see the surface structures that delineate a part of the body (for example, the eyes) as composed of many dynamically interacting sub-structures (retina, blood vessels, tissue, cells, etc). Exercise 4 extends such perception to the entire body and its networks of dynamic interactions, amplifying the dynamism so that such interactions become"translucent." In Exercise 5, we are asked to go further—to open up the"outlines" of every surface and structure of the body until they are released to Space.

These exercises activate a penetrating way of knowing that 'sees' objects in a new light. Upon closer analytic scrutiny we see that any form that we normally designate or single out as a solid object can also be seen as a dynamic system of interdependent interactions. Whether we refer to a whole (e.g., the body), or a part (e.g., the eye), no object can be isolated or seen to exist on its own. What we commonly perceive as 'our body' has no intrinsic, self-existent reality; rather, it is a composite phenomenon, as is each so-called part. The more we exercise and become familiar with this more penetrating kind of knowing, the more we begin to appreciate that the body (and all appearances of form) are malleable and lacking solid substance.

As we learn how to counteract our tendency to fixate on objects (in this case our body) as solid and opaque things, a new type of "knowing faculty" emerges. Tarthang Tulku suggests that this amounts to a "mode of seeing not limited to a particular position or point of view" (p.27). In other words, this new knowing faculty is not derived from taking a stance, a position set apart from the object to be known. Nor is this new mode of seeing predicated on the conventional elements of knower, knowing, and known (or perceiver, perceiving, and perceived). Instead, we can imagine it as pervading an "open field or dimension" (p. 28), which is already 'in' experience 'before' experience is divided into conventional dualistic categories.

In the latter part of Exercise 5, the form of the body is released into Space until it is completely gone (and "gone beyond gone"). This experience elicits the insight that the ultimate or true nature of any object or thing—any feature of ordinary appearance—is fundamentally Space. As Tarthang Tulku states: "So perhaps 'things' could actually be said to be, fundamentally, nothing but this 'space' or 'openness'." (p. 30) Yet Space is not merely empty, devoid of qualities: the exercise also suggests that a new knowing faculty is co-emergent with a greater recognition of Space. The 'knowingness' that emerges restores our capacity for "primordially fresh encounters," clarity, and ease. We might say that such 'knowingness '(what the text later calls Great Knowledge) abides within the nature of Great Space (which is not a 'thing'), yet can still cognize the appearance of conventional things 'in' lower-level space.

This latter point deserves further comment. Great Space is not an object which can be known by a subject. In other words, Great Space is not an experience to be grasped at or attained. And because Great Space is not an object of (ordinary) cognition, it is beyond thought and our attempts at grasping it in a conceptual manner. On the surface this might suggest that the different way of knowing the Space exercises are geared toward helping us access is either inaccessible or of little value to our life. However, because Great Space is infinitely open, it does not exclude thinking or thoughts. It is just that the path to Great Space is not traversed by staying within the dualistic limitations of lower-level space, time, and knowledge (such as substance-body/linear time/ordinary thinking mind). We might liken a 'Great Space-Time-Knowing' to a completely uncontrived naturalness and undirected awareness, or what the TSK text refers to in a special sense as "intimacy." Part of the benefit of the TSK exercises is that we learn how to differentiate between lower-level knowledge and Great

Knowledge and subsequently become more familiar and adept at relying on the latter.

Great Space is not actually elusive, remote or mysterious. Rather, our usual self-orientation (which in TSK is equated with an active lapse in knowing) gets in the way of appreciating that every condition, every event, every thought, form, or circumstance is expressive of Great Space directly. Great Space is not something that we are missing, nor something to be seen or experienced in contrast to ordinary appearance. It is so close (an "uncontrived intimacy"), that we fail to know it, and thus fail to know the true nature of appearance as it is. Non-localizable, Great Space is already 'here'. It is just that as long as we are intent on grasping for an experience or object, we tend to look elsewhere.

There are many adjectives one could evoke to express the nature of Great Space: ungraspable, inconceivable, unlocatable, unapproachable, unconditioned, unborn, baseless, traceless, immeasurable, limitless, boundless, boundaryless, infinite, transparent, open, free, deep, and empty. But words are only crude pointers. We could imagine Great Space metaphorically as a single sphere, non-dual, unbroken or whole in the sense it is never divided or compartmentalized into separate existents. Great Space (seen with Great Knowledge) does not affirm the arising of any form or existent (or non-existent), since it is an inexhaustible absence of obstructive phenomena.

TSK exercises help us unlearn, relax, and open up the tendency toward grasping for experience. The exercises described so far help us get a feeling or appreciation for

Great Space by relaxing our fixations on ordinary appearance, particularly our taken-for-granted sense of being an isolated subjective knower engaged in experiencing a substantial external world. Exercises 4 and 5 help us do this by encouraging us to see the body more in terms of Space. Yet to stop at this point would lead to the erroneous conclusion that it takes a special or extraordinary experience to perceive the body (and all appearances of forms) as essentially Space.

The transition to Exercise 6 in the Giant Body series challenges this assumption, inviting us to critically examine and compare the ordinary appearance of the body as dense and opaque with the newfound translucent way of seeing of the earlier exercises. Tarthang Tulku poses the question, "Is it possible that both of these versions, as they are experienced directly, are equally a kind of 'space' experience?" Underlying the question is the suggestion that there is no real dichotomy between appearance as substance and the accommodating nature of Space. That is, what appears as unquestionably dense and opaque is just that: an appearance that is inseparable from the nature of Space.

Space is thus not remote. It is not some hidden absolute or underlying stratum of reality. Rather, the union of appearance and space is felt as an embodied quality of experience.

A more formidable question now emerges: Given the appearance of form-body, is there really any thing substantially there? If we are challenged to look for or find 'the body' or any part of the body, what will be the result of our analysis? As Exercises 1–5 demonstrate, our analysis first yields a deconstructive de-layering of the body and its parts. This transitions to a complete openness devoid of substance. The form of the body is seen to be a conventional frozen perception (in TSK language, an output of a particular 'focal setting'), a mere label which is imputed upon a dynamic interaction of composite forms which themselves have no ultimate point that cannot also be opened into Space. Exercise 5 challenges any lingering sentiment that the body inherently exist as substance, while Exercise 6 ("Opacity versus Translucency") challenges us not to go to the other extreme that "nothing exists." The appearance of the body as form is not being rejected; what is at issue is our habitual tendency to view/experience the body as something solid, rigid, and separate from space.

The Commentary to Exercise 6 is very instructive, as it suggests that there is an intrinsic unity between appearance (forms) and Space, and that we have the knowing capacity to keep this unity in view. It is here that Tarthang Tulku begins to utilize such phrases as "higher-order space" and "higher-order knowledge." As he states (p. 30): "We might say that higher-order knowledge does attend to conventional items and perspectives. However, it does so in a way that appreciates their participation in the higher-order space." My interpretation of this statement is that we can access a higher-order knowing capacity that can operate on several levels simultaneously. Knowingness is cognizant of appearance (forms, bodies, objects, mind, events, etc.) as fundamentally Space, while simultaneously cognizant of the conventional features of appearance in all their multiplicity and complexity.

Practice of the Giant Body series up to this point lets us glimpse a new view of reality and inspires a new appreciation of a higher intelligence. Experientially we begin to feel a greater sense of openness and ease. We gain familiarity with a knowing capacity that can remain aware of the fundamental nature of appearance as completely open and space-like, while still being able to operate in a field of conventional designations habitually predicated on a set of dualistic concepts (self vs. world, subject vs. object, body vs. mind, existence vs. non-existence).

With the experience of space more available within ordinary appearance, we are less taken in by what ordinarily would seem solid, real, rigid, and opaque. Situations we encounter seem more flexible, fluid and workable. As we relax more, we are less caught up in our habitual, conceptual way of reacting to conventional appearance. Less physically tense, less emotionally reactive, less mentally blocked, we discover more options for how we can 'be' in any situation. We get a taste for the totally unobstructed quality and freedom from restrictions that 'higher spaces' offer.

However, this statement has to be qualified. While brief glimpses of "higher spaces" are a gateway into the vision, fruition requires stabilizing and integrating 'space experience' with the apparent movement of time. And the Knowledge exercises later in the text are designed to peel away the gross and subtle layers of dualistic habits associated with conventional 'lower knowledge' that ordinarily obscure the more open, space-like qualities of 'higher knowledge.'

## Transition: No Freezing Tendency

Sincere and dedicated practice of Exercises 1–6 will generate inspiration to go deeper into the remaining Space exercises. Preliminary to the next series of exercises, the text investigates a model for the development of a person in a world (pp. 32–34). Four alternative views are presented, suggestive of a developmental process for the emergence of our ordinary worldview that presupposes the birth (and death) of a person with a permanent identity in a standard world order. I have found myself going back to contemplate this section many times over the years, and it serves as a good example of how one can gain access to different levels of meaning of the text over time.

Contemplating the meaning of the four-stage progressive model can help to loosen our habitual tendencies involved in 'freezing' space and time—giving rise to the appearance of opaque and stable things that also appear to undergo change through time. Tarthang Tulku states (p. 34): ". . . persons and objects . . . are only lower level summaries of the ongoing tendency toward freezing and screening out the full openness of space."

The progressive model suggests that such appearances in space and time are dynamic rather than static, and that any reference to a point of permanence or stability is the result of some sort of "freezing tendency." Our ordinary notion of personal identity is an example of such a habitual and ongoing tendency. Yet as the fourth stage of the model puts it, even the process of freezing is also space and time, and has no real or inherent existence that sets itself up in contrast to the freedom of space:

> No 'freezing' or anomalous tendency has set itself up in contrast to 'space'. The frozen pattern is neither frozen nor even a 'freezing'. Worlds, things, and persons remain 'space', rather than only deriving from 'space' as in [stage] '3'. (p.34)

Internalizing this view can help us to further relax. More familiarity with the immediate availability of space can help to facilitate practice of the exercises. We need not try so hard or exert so much effort, as the opposition between space and objects is seen to be only a stubborn pattern rather than the way things are. While practicing the exercises, we can gain more confidence in this view by keeping in mind that we have never really strayed from Great Space.

While it seems in ordinary terms that 'we' are embarking on a particular exercise, Great Space is unconditioned, and doing the exercise is not going to improve Great Space, create it, or bring it about. In other words, we do not need to meddle so much. If we view the practice as the cause for bringing Great Space into view, we are still practicing from a 'first-level' focal-setting, where we see the 'self-as-agent' doing the practice. We will then be overcome by an incessant worry whether we should be doing something or not doing something or whether we are doing the practice correctly. We will have too many expectations or fears or concern ourselves too much with whether we will attain or lose some special experience.

TSK practice involves progressively unlearning the tendency toward seeking. We do not realize or know Great Space by moving away from some place or status that we consider to be lacking in this open dimension toward a distant state or condition that is more characteristic of Great Space. That is often the mode we start from, propelled by a strong desire or goal-orientation to attain a more desirable state we presume is not available in the present moment. In some respects, the TSK exercises offer training in the confidence that will allow and deepen our experience of what already is. They invite a profound non-doing that allows Great Space to permeate our preconceptions of doing. We might think of allowing the 'Grace of Great Space' (my expression) to facilitate the unfolding of the exercises, rather than being directed by the 'effort-driven I'.

If we can allow a relaxed but alert orientation while 'doing' the exercises, we can come to see directly that the body's real nature is space. This assertion is not merely a philosophical musing or metaphysical sleight of hand. In fact, conducting the exercises protects us from deadening embodied insights into mere intellectual concepts 'about' space. It is quite easy to engage in 'TSK-speak', reducing space to just another mental concept. Simply to believe and espouse that "everything is Space" is to hold in mind a concept. In that case the limits of conceptual mind remain solidly entrenched, obscuring the true nature of Space.

These considerations apply to the person as well as the body. The essence of a person is not a permanent identity or a body, but what Tarthang Tulku refers to as "a condition of open uncommittedness" that appears to become frozen or "consolidated," collapsing into a standard three-dimensional world order. Like other discrete entities, persons are subject to the laws of causality. They are born, remain relatively permanent and unchanging through time, and then suddenly cease to exist at the moment of death. In this way, existence/non-existence is translated into life and death. But this view presupposes that personal identity exists in some atemporal position as a bystander to the linear flow of time (until it encounters its apparent destruction at death). Our attempts to freeze space are closely linked to the desire to maintain the permanence of this self. The result is that we inflict on ourselves an inherent sense of lack and dissatisfaction. By challenging the tendency to freeze and consolidate space, the Giant Body Exercises can have a powerful effect on the felt experience of being alive.

## The Embodied Mind Exercises

The next set of Space exercises (7–10) investigate the embodied mind—the lived dimensions of thoughts, feelings, sensations, volitions, and the conventional partitions between body and mind, as well as the emergent position of the observer. This exploration is critical to the TSK vision, and deserves extended commentary.

At first glance, Exercise 7, "Body-Mind-Thought Interplay" resembles mindfulness practice as taught in various Buddhist traditions. But there is an important twist to the exercise: we are asked to note how various interplays of body-mind-thought constitute "you in that situation" (p.35). This meta-observation helps us to take notice that an observer is constituted in juxtaposition to the stream of mental events, an observation amplified in Exercise 9. We tend to take the existence of this observer for granted, perceiving that we (as observers) are the owners of our thoughts, feelings, and experiences. But for this to be the case, partitions must exist between the observer and the constituents of experience. We ordinarily say, "I have a feeling, I am thinking a thought, I had this memory, I have a headache," as if the observer were separate from the body, thoughts, feelings, volitions, and perceptions. Such a view presupposes the 'self-as-doer', 'self-as-agent', 'self-as-experiencer', self-as-knower'. Exercise 8 challenges us to examine these partitions and open them up to space, just as we opened up in the Giant Body exercises the outlines that define particular components of our body.

Practice of Exercise 8 also sensitizes us to how our mind perceives and then reacts to its own perceptions. Not only does the observer stake its claim as an autonomous agent, producer and owner of its mind, body, thoughts, feelings and actions, but it also identifies with them. There seems to be a contradiction. On the one hand the observer claims to be positioned in some aspatial and atemporal location—the seeming passive watcher. Yet usually we find ourselves completely identified with our experiences: they are mine, and I am them. The more we observe, the more we may be astonished to discover the extent to which we are enmeshed in self-created mental prisons, built by the recycling and consumption of habitual reactive patterns. In a constant flow, information received from our senses fuses with our prior memories, concepts, and past habitual reactions. Like our belief in a world independent of our perceptions, our commitment to a reified flow of thoughts, feelings, and perceptions amounts to a kind of modern idolatry.

Exercise 8 skillfully guides us to embrace the interplay of experience along with the partitions that normally constitute our feeling of being separate from that experience. It asks us to offer it all to Space. I like to think of Exercise 8 as an 'inner Space massage' that dissolves the seeming separation between mind-body-thought-feeling.

Practicing Exercise 8 can help heal the sense of fragmentation, dividedness and inner conflict we often experience between our body-mind-thoughts-feelings. We have grown used to viewing the mind as frozen and compartmentalized (Just think of the way that the Myers-Brigg Type Indicator, a popular psychological assessment instrument, relies on polarities such as "extroversion/introversion," "thinking/feeling," "intuition/ sensation," and "judgment-orientation/perception-orientation.") Opening up the body-mind complex not only results in integration, but frees up the energy frozen into 'mental blocks', habitual defensive postures, and inner conflicts.

Exercises 9 ("Participation as Observer; Participation as Embodied Person") and 10 ("Participation and Space") are more inclusive than Exercise 8, because they take into account the subtle position of the observer, co-emerging with the perception of objects. These exercises give guidance on how to completely open up our experience of the mind-body-thought interplay, with a special emphasis on letting go of our notion that the method being employed is a cause for transcendence. The result is "unqualified openness" that does not require an observer in order to be 'known'. We could call this an 'auto-cognizant space awareness', a way of knowing "prior" to subject and object and mental commentary of the observer.

Exercises 9 and 10 help us to see the self as a habitual configuration of space-time-energy that has become frozen. As Tarthang Tulku notes (p. 18), "the self exist in a frozen energy state," and "thus we are not flowing-floating properly" (p.18). These habitual configurations impede our natural freedom, blocking our creative capacities, while inhibiting our ability to respond fully to the challenges and changes we face.

In practicing these exercises, our habitual tendency to 'know things' makes us 'come out' of such a non-dual mode of awareness. We are asked to pay special attention to this "coming out tendency" (which may take the form of trying to report back to oneself, or grasping the non-graspable nature of such an unqualified openness). The "coming out tendency" reveals the habitual belief that the observer is a real, self-existent entity—and that knowledge can only arise through the position of the observer. Practice of these exercises reveals on the contrary that the observer is a lived presupposition, a participatory fiction. By freezing the open dimension, it collapses space into a locatable and fixed position, which the observer then claims to 'occupy'. This occupation (or preoccupation) lies at the root of a self-orientation, the founding story "I am." Biased, distorted, and delusional, it obstructs Great Knowledge. Just as there is no inherent existence to the Giant Body (or its constituent parts), there is no inherent existence of the observer and no basis for asserting such existence.

Knowing without being filtered, channeled down or biased by an observer may seem unfathomable. We are so deeply conditioned by our conventional dualistic view that it is hard for us to imagine any alternative. But consider a thought experiment. Entertain the possibility of awareness floating free in the emptiness of deep space, with no objects in sight. Lacking any solid points of self-reference or external reference, awareness would have no need to contract or fixate on some point in order to reflect on itself. Knowing would simply be free to know, resting naturally, abiding in the emptiness of free-floating space, without any tendency to 'come out'. If reference points did appear or emerge, they would be seen as expressions or symbols of deep space, characterized by a kind of "zero gravity" (DTS). Thoughts generated from the cognizing of these substance-less appearances would be 'taken lightly' as symbols of Great Openness. The imaginary zero-gravity world would have no hold on Great Knowledge, and perception, never departing from Great Space, would be freed upon arising.

In contrast, our ordinary knowing could be thought of as arising within a single sphere, a beautiful crystal ball. Mistaking the reflections visible on the inner walls of this great sphere as an external world, made up of objects 'in' space and time, we come to view ourselves as observers at the center of the ball, viewing an environment or external world 'out there'. The only way for the observer to make sense of what is observed is to 'come out' of what is observed, singling out forms and phenomena and referring them back to the position of the stationary observer. Both the observer and the objects it perceives are reified into self-existent, solid entities, *occupying* space rather than being space. It is as if the crystal ball repeatedly shattered and cracked apart into the fundamental duality of 'I' versus 'the world'. In this implosion of knowing, the gravity of identity prevails, sucking awareness into a black hole occupied by the observer. The 'I' (one's 'own mind') is felt to be the source of all knowledge and experience, while space fills up with finite objects.

Such crowding and proliferation lead to a kind of endarkenment that shuts out higher spaces. Cut off from the light of awareness, our bodymind complex chases after the reflections that arise in the field of this lower-level space. Since each reflection is insubstantial and transitory, no reflection can provide lasting fulfillment or satisfaction. Fixated on the knowledge and concerns generated within this field, we lack clarity, energy, higher purpose, and joy. We construe ourselves as the 'field-assemblers' of the 'field communiqué' (DTS), not recognizing how knowledge has degenerated, muddled and confused by the endless processing of our own secondary reflections. In the dark field of the shattered crystal, friction and disharmony prevail. Forms and perceptions appear solid, feel heavy, and are taken to be unquestionably real.

The TSK book uses a similar image: a confining chamber in which echoes (meanings) reverberate off the walls, and even constitute the walls (pp. 102–3):

> In returning to the lower space images of a platform or vantage point for observation, and the image of a containing chamber, we can see that it is precisely meanings that, in order to mean, require a relatively static position defined by similarly static surroundings (and container). Meanings deal in terms of 'things meant', which must necessarily be marked out by unbreached boundaries, opaque partitions. Meanings depend on limits being set on our attention to the dynamism and process of partitioning.

Focusing exclusively on substance and content, the observer comes out and declares its independent existence. We accept as a given that some independent agent is the producer and receiver of our thoughts, feelings, and perceptions—that the events in the external world are happening to an internal, isolated, but constant point of reference, the 'me'. Our lived view is that "I have a body," "I have a mind and experiences," "I am subject to constant change," "I am born," and "I will die." The fundamental duality between subject and object correlates with a mental activity or thought-generating process that is intent on 'pointing things out', fixating on meanings derived from linguistic and conceptual structures. With all of our attention fixated on the rational, linguistic stream of consciousness, no other alternatives for knowing seem availablc.

But what if the observer in the crystal is only another reflection? What or who would 'observe' the observer? Who would 'own' knowledge? Or do these questions as such still reflect the need to force knowledge into a dualistic structure?

Consider the analogy of dreaming. The dream is structured into a mosaic of the dreamer, dream characters, and the dream setting or environment. Yet all of these components are a unity within our consciousness. There is no inherent, independently existent subject or object within the dream. Usually the dream is felt to be real until the moment we wake up, so that even here the observer is unobserved. Yet if we wake up within the dream (lucid dreaming), we have the chance to realize that even the observer (or dreamer) is insubstantial, not inherently real. Knowing this, we feel tremendous freedom. It is not that we suddenly disappear or go blank. Rather, we feel liberated from the drama of the dream. We can play as we wish.

In Exercises 9 and 10, the TSK text presents the nature of the observer as co-emergent with the focal setting on Space. When the observer 'comes out', it is like falling back in the dream of duality. But can we nourish our insight into the 'duality dream' by viewing the bodymind complex (which includes the sense of a bystander-observer) as an illusory image, an image that appears to be solid, fixed, and separate from an external world? Rational analysis can help: if all I perceive is a projection of meanings, if no objects within my range of perception exist independently, 'from their own side', then this 'I', the seeming organizer of this field of experience, may also be a dependently originated construct, void of an inherent, independent existence. The apparent sense of being a 'by-stander' to our experience can be recognized as a limited space, one that can be opened up.

Yet let us not get too inflated; Exercise 10, which explores these possibilities in some depth, still offers only a glimpse of Great Space, a 'transitive view'. We can understand this qualification on several levels. First, insight into the illusory nature of the observer can also become a new mode of grasping—taking hold of the absence of an observer! Rather than declaring victory, we must turn to investigate the voice that claims it cannot find the observer. Does the mind that is apparently opening up the observer have any inherent or independent existence of its own? Is there really some independent entity called 'the mind' that is doing the exercise, or is this 'perception' also the product of the focal-setting? (Exercises 11–13 will take on this challenge directly.)

Second, the insight that challenges the observer is still couched within a hierarchical framework (first-level, second-level, and third-level space) that suggests that the 'place' we start from seems temporally and spatially distant from the space and time of higher realization. To try to 'get rid of the observer' or fight 'the ego' from this perspective may end up affirming the realness of such designations rather than relaxing and seeing through them.

Third, experiences will quickly fade unless practice is sustained. Glimpses of higher spaces need to be 'achieved' many times to counteract the coming-out tendency. Having an intellectual grasp of higher spaces based on a few partial experiences is a far cry from the knowledge that comes from having done the exercises concertedly over time. Talking about or intellectualizing TSK material is not the same as actualizing insights through embodied experience.

## The Openness of Space as a Challenge to the Self

In many respects, Exercise 10 seems to be a pivotal point in the text. It also forms the foundation for Exercise 14, "A New Focus on Space," which leads to a more experiential insight. But I have found that it is just at this point that habitual tendencies toward grasping raise their ugly head. Our cherished self-image—that illusory observer that we have built our entire life around—does not surrender its position so easily. Deep insight into the insubstantial nature of the entire body-mind complex, along with the so-called external world, is not an easy pill for the self to swallow.

Exercise 10 calls into question the image of the self occupying a 'change-free' zone that leaves it both continuous and independent. But rather than accepting that the observer can be understood as a distorted or biased version of Great Space, the self prefers to hold on to its existence, even if this means remaining sick and deluded. Taking up a defensive posture, the self attempts to freeze space and time. At a subtle level, insight into the absence of an observer can also become frozen, a subtle 'coming out' that sets another cognitive obscuration in place. Adherence to such an insight into Space becomes yet another position for the self to proclaim, a clever dogma and an opportunity for indulging in 'TSK-speak'

Exercise 10 (and also Exercise 9) anticipate this response by challenging us to let go not only of our fixed ideas of self, but also our preconceived notions of the absence of the self/observer. Simply quieting the mind and watching the body-mind-thought interplay is not enough. When we are instructed to open everything up in the body-mind-thought interplay (including the observer) until it is all space, even our notions of 'opening up' and 'being gone' have to be opened up. This amounts to a letting go of all conceptions and views, a complete surrender of effort and habitual tendencies of needing to fix, improve, change or alter our immediate experience. In a very real sense, there is nothing for the self to do.

In the TSK vision, the self's stubbornness and defensiveness is still Space, Time and Knowledge. However, from the position of the confused, illusory observer, no easy reconciliation is possible. The prospect of letting go into the freedom and liberation of higher spaces feels like a great loss, an existential terror, a complete dissolution into non-existence. At this point we may back off from the TSK exercises. We may tell ourselves that we no longer have time to practice; or we may keep a safe distance by restricting our involvement to reading the text. We may confuse intellectualizing our understanding with actualizing it, or fall into doubt, or seek out a different method/teaching and start the whole process all over again. Or we may simply give up.

That is why honesty and perseverance are necessary to actualize this great vision, and why practicing with other serious students can be such a great support. The TSK teaching is not just a compendium of philosophical views, but an experiential vision for liberating the capacity to be fully human. This may sound attractive, but in practice it can lead to challenges we are not prepared to face.

Exercise 10 seems relatively simple, amounting to a letting go of effort. Yet it is incredibly difficult for the self to do, because the self is itself an obscuration or distortion on Great Space. The exercise calls for shifting from a potent self-as-agent view—the felt sense that I am engaged in effortfully doing the practice—to a more open, allowing, and gentle unfolding of experience. The self's existential anxiety is due to the mistaken notion that such openness will annihilate existence, that the self will disappear into nothingness. Caught in the duality between 'is' and 'is not', the self is fearful that unless it remains the center of attention, it will not be able to manage and control its world. As Tarthang Tulku states in TSK, "Great Space is too much for the self to handle." And so the self freezes experience, affirming its realness, permanence, and independence at the cost of enormous expenditures of space-time energy.

The lived view, or output of this focal-setting is that consciousness is located inside one's head, 'behind one's eyeballs', looking at discrete objects, independent 'things' that are separated by space. We interact with what appear to be solid things 'out there' by moving from here to there. This mode of awareness typifies lower space and time, a designation used in TSK to signify a limitation on our knowledge and way of being. Awareness becomes 'desire-directed', driven to grasp fulfillment by acquiring 'things' or experiences that seem remote, distant, or hard to get. As the 'I am' is afflicted with an insatiable feeling of lack, the nature of mental experience is one of constant movement or unsettledness. As our awareness is 'taken over' by continuous thought-streams of identification (manifested as 'hope and fear stories'), we lose sight of Space and Time as resources for human being.

In this respect, the Time, Space, and Knowledge vision is really an active inquiry into and process of recognizing our true identity, the nature of our own being. Is the self with all of its anxieties, limitations, and fears our true identity, or can we look instead to Great Space, Great Time, and Great Knowledge? This is the central question, the heart of the matter that activates inquiry and embodies the vision.

Concerted practice of Exercises 9 and 10, along with 'thinking things through' offers the higher-knowledge insight that the demarcations separating subject and object are simply the way 'reality' appears when seen through the lens of the observer or bystander. With this insight, we begin to 'see' appearance as appearance (rather than as being absolutely real). Even 'solid', 'concrete', and 'inherently existent' are more appearances. When we know on a higher level that we are really seeing 'no things', our need to fixate or grasp is released into the freedom of space.

Appearance in Great Space is not compartmentalized into such constrictive solid entities. Rather, Great Space is co-emergent with a knowing dimension (knowingness or Great Knowledge) that is beyond the realm of an isolated subject knowing a distant object. This seeing is the result of a 'thawing out' process, an unfreezing of the bodymind complex that serves as a fitting metaphor for this whole series of Space exercises. It is as if we learn how to melt the solid iceberg of our frozen body-mind complex until it merges with the deep and vast ocean of Great Space. Of course, the frozen ice and deep ocean have the same nature. We are shocked to see that there was never any solid or self-existent body or mind there in the first place.

## Shifting the Center of Gravity: From Self to Great Space

The Space series of exercises in TSK could be condensed into one question: What would it be like to be full of Space rather than full of oneself? Rather than feeling and thinking that experience was happening to an innate, isolated self, what if we were able to experience experience as simply 'happening', arising as Space, in Space, to Space? What would such a radical re-ordering of perceptual embodiment imply?

To put Space at the center is to be in the mandala of the body-mind complex. 'We' arrive at the 'centerless center', where all forms, all perceptions, all sensory experiences are directly Great Space. Putting Space at the center silences mental commentary and narration. At the same time, it allows direct sensory perception and even thoughts unmediated or biased by an observer. Free of the gravity generated by an insecure self, we participate freely and joyously in the intimate dance of Space and Time.

Space-centered concerns are wider, more comprehensive and inclusive in scope than self-centered concerns: "one point is all points." Instead of an isolated observer weighed down by the gravity of identity, a dynamic presence embodies the whole. Secure in the fullness of Space, our conduct is no longer compelled by self-centered concerns (Etymologically, "security" comes from the Latin for "carefree.") Fear lessens and the heart opens, for in this centerless center every appearance is united. As mind and heart are opened beyond the boundaries of 'here and there', 'me' and 'mine', genuine and spontaneous responsiveness wells forth. Being 'Space-Full' but 'self-less', we emanate Great Love, spreading Great Knowledge in all that we think, say and do.

# Space Projecting Space

Jack Petranker

*When we try to examine the mirror in itself we discover in the end nothing but the things upon it. If we want to grasp the things we finally get hold of nothing but the mirror. This, in the most general terms, is the history of knowledge.*— Friedrich Nietzsche

First-time readers of Tarthang Tulku's *Time, Space, and Knowledge* are often brought up short early on by a statement in capital letters on page 10 of the book. Here it is, with a bit of the preceding text:

> It almost seems that an object which is 'here' has appeared from a past condition of empty space, needs that space to be here, and gives way to a future condition of, again, empty space—'non-existence'. It looks like space is projecting space—that is all! Even what seems to fill up the

space as a present tense, existing object, is also space. So,

SPACE IS PROJECTING SPACE INTO SPACE!

In context, this statement seems to say that our entire reality is nothing but space and the 'activity' of space. Most people do not know what to make of this, and the mystery is deeply unsettling. I have even met a few people who told me that when they came to this sentence, they felt utterly and completely stuck, and put the book down.

I want to explore one way of investigating the meaning of this challenging statement. I do not claim it is the 'right' way, or even that it remains true to the discussion that follows the quoted phrase in the original text. But I do think it can give a sense of how space actively shapes our experience, a fundamental insight at which the quoted passage is pointing.

## I

What is space? The straightforward view is that space is what contains physical objects. But the TSK vision draws out in countless ways a more encompassing understanding, one that can be expressed by saying that space is the necessary accommodating matrix for whatever we experience.

These two very different understandings of space—one focused on objects and the other on experience—can be linked. Ordinarily we think of an object as something independent of the one who experiences it. But objects appear to us as independent only *after* we have abstracted their identity from the experience in which the object appears to us. After all, by definition the objects we actually have contact with are not independent: objects as we know them are precisely objects as we know them. We can theorize about objects all we want, but we can only know what we know, only experience what we experience.

Whatever we experience—whether it be a thing (a chair, a tree, the jacket we put on when we feel cold), a memory, a feeling, or a sensation—requires a space in which to appear. You might wonder if this is really so, or whether 'space' is just an abstraction we use to make sense of the possibility of experiencing. But if you look into the threefold structure 'I-experience-something', you can see at once that interrelationship is essential to all experience. For interrelationship to happen, there has to be a linking and a matrix, and that linking matrix is space. That is why it makes sense to say that the space that contains objects is also, and more fundamentally, space that accommodates experience.

Understanding the link between space and experience helps us see that space comes in many varieties, each of which has the capacity to accommodate particular kinds of experience. The perception of physical objects, along with the physical objects perceived, is one kind of experience and depends on one kind of space. Other kinds of experience (for instance, memories) depend on other kinds of space. What shows up in one space may show up in a very different form in another space, or may be barred from showing up at all. For instance, scenes from the past that regularly arise in the space of memory do not show up in physical space. If they do, we call them delusions or hallucinations, expressing by these terms the basic assumption that the different spaces appropriate to different kinds of experiences are incompatible with one another.

Suppose I am walking down the street. Brilliant roses bloom in the garden on my right; my new running shoes put a pleasing spring in my step; the wind streams past my face. But after a while, I start to think of what I will have for lunch. Now I am somewhere else, planning a menu or a shopping trip. As this very different kind of experience wells up, I do not just shift the object of my attention within a particular space (for instance, from perception of wind to perception of running shoes): I shift spaces. The space of the meal I will be having in a few hours accommodates in a particular, restricted way the supermarket where I will buy the food, but it does not does not accommodate at all the wind on my face. And the space of my walk down the street does not accommodate the anticipated meal.

Tied directly to the various and constantly shifting kinds of experiences available to me, every shifting space I inhabit is given structure by a founding story that lets me make sense of it. (for more on the founding story, see LOK, parts 2–3). This sense-making structure means that for every space, I could in principle give some account of my presence and participation in that 'worlded' space, including the various ways I relate to the other entities that the space accommodates.

Each space we inhabit is thus inseparable from the knowledge that tells us both 'what's happening' and 'the way things are' within that space—the knowledge that sets up the prevailing story. We do not first enter a space and then make sense of it; rather, we can shift to the space that will accommodate a particular kind of experience only as a sense-making, story-telling knowing is activated. Without this knowledge, we cannot inhabit space. It comes with the territory.

The intimate relation between space and knowledge gives us two distinct ways of looking at space. Because it has the capacity to accommodate unlimited appearances (countless stories), we can say of space that it is empty. But because it accommodates only what makes sense within a specific known world of experience, we can also say that space is full—filled to capacity by the world that the story makes available. This filled-to-capacity space is in turn full of the specific objects it can accommodate.

This twofold space-structure lets us reframe the statement that "space projects space into space." Space as shaped by the 'story of what is so' projects that story into the limitless capacity of space as empty. Within that story, specific objects are projected, and they too are space, in the sense that their arising depends completely on the space within which they arise. There we have it: space projects space into space!

Space-projecting typically passes us by, because we are too focused on objects to notice the space that accommodates, sustains, and even projects them. As a result, we are also blind to the knowledge that informs space, the knowledge that 'worlds' space, even though we rely on that knowledge completely. The TSK vision of reality gives access to that inherent knowledge, belonging not to us, but to space itself. The more we exercise this unexpected capacity for knowledge, the more the invitation to see all appearance as "space projecting space" takes on force.

## II

I turn now to some practical suggestions for how to activate the space-knowledge toward which the "space-projecting-space" insight points. I will present these suggestions in terms of three (overlapping) kinds of experience: the perception of objects, the felt experience of our own embodiment, and the world of the mind, populated with thoughts, emotions, moods, and so forth.

### A. The Perception of Objects

Perhaps because we link space with the physical realm, it is fairly easy to get a sense of 'object- centered' space. Look around the room, noticing the objects that meet your gaze. Be sensitive to the space that 'contains' those objects. This will include the space that surrounds the object, the space that the object occupies, and the space that the object shares with other objects (along with the sense of that shared connection). Cultivating a sensitivity to these aspects of space does not mean turning away from the objects that space reveals. Nor does it amount to the claim that these objects are 'nothing but' space. The objects are what they are, appearing along with their accommodating space.

Experimenting with the space-openness of objects in this way seems to make objects less weighty or heavy and more light or transparent. They become more fluid in their identity as well, for the space that accommodates two or more objects does not really differentiate between them. Experience itself may feel lighter. All of these tendencies are strengthened if you consider yourself (not your body, but your 'self') as another object occupying the same space as the physical objects you perceive.

Another way to experiment with the space of the perceived world is to move around in that world. As you walk through space, it may feel as though you are carrying the space you occupy with you (see KTS, ch. 29). At the same time, however, you are entering a new space with each step. Asking in terms of your experience how these two ways of experiencing space interact helps loosen the tendency to turn space itself into an object, which can otherwise be a limiting factor in the exploration.

Cultivating a sense of space-lightness and openness 'within' the objects we perceive can be both relaxing and illuminating, but eventually the tendency to focus on objects over space will reassert itself. Even if objects continue to seem lighter and more accommodating, this may simply mean that we have entered a different space. In this new space, perceived things may manifest a different and less burdensome 'weightiness', but this is only a first approximation of the "space projecting space into space" insight.

One way to deepen this emerging insight is to investigate the knowledge dimension of space. When an object appears in space, it bears an identity, which in turn is sustained by a story. *Love of Knowledge* explores the "here I am" at the heart of this story: the self asserting its own identity (the 'founding story') in relation to the world in which it operates. Story and identity shape every encounter between self and the physical world it inhabits, and it is in this interaction that space-knowledge makes itself available.

For instance, as I type these words, I can be aware of the keyboard, my fingers, the computer screen, and so forth. I can also be aware of the space these objects share, and of the way that this space, by linking each of these objects with the others, challenges (in a felt sense) the boundaries of fixed identities. Beyond that, I can be aware of the knowledge-bearing space within which all of this makes sense: the 'world' in which I am writing an article, anticipating the reaction of its potential audience, reflecting back on how the article has been taking shape, and so on. It is out of this 'worlded' space, with its inherent coherence and meaningfulness, that each new moment of experience emerges. The more I can contact this unifying whole (without losing contact with the immediacy of the particulars), the more appearance is empowered as an expression of knowledge.

Notice that it is possible to expand space-experience in this way without an explicit focus on space. All

I need to do is ask after the larger meaning or context of what I am doing right now. Another approach is to simply let the object be in the wider web of its own relations, without losing track of my own connection to it. Doing this is something like letting two spaces interact, so that their inherent spatiality and accommodating capacity stand out.

An obstacle to this kind of inquiry is 'losing track' of objects, which also means losing track of space. When we first focus on the space within which objects are embedded, we may find the resulting experiences deeply interesting, so that we naturally keep our focus. But after a while, the old patterns that focus on objects rather than their accommodating space emerge. We turn whatever experiences we have been having into objects in their own right, and this shift takes us back into the standard mode of experiencing, in which we occupy space without being aware of space. Having settled back into this old habit, experience loses its special aliveness, and soon the mind has wandered away from the object of perception and inquiry. Lost in space, we drift unaware from one space to another.

One way to counter this tendency is to continue to explore the space dimension of experience even when our attention shifts away from the original focus of inquiry. Whether experience presents physical things and settings, remembered situations, ideas, worries, or anything else, the space that accommodates that presentation will be available. Since that is so, 'losing track' of the space of physical objects need not mean losing track of space. In fact, non-physical space offers interesting complexities that can deepen exploration. For instance, when we remember something that happened to us recently, there is the space of the remembered scene, including the space-relations between various objects in that scene; but there is also the space that situates that memory alongside other, related memories, as well as the more encompassing mental space within which the memory as such arises.

In carrying out such investigations, it generally seems better not to focus on space directly, but instead to let space be available in the course of being aware of objects in space. To focus on something directly is to make it into an object, and the moment that space becomes an object, it can no longer serve to accommodate. (A possible exception is when one is aware of an object in another space while inhabiting a more encompassing space; for example, in viewing a painting.) Instead of making space the focus, it seems enough to let objects be 'space embedded'.

## B. The Felt Experience of Embodiment

The space of physical objects is also the space that we inhabit in being embodied. Our embodiment, however, is not physical in quite same way as everything else, because it has a felt dimension to it. For instance, when I rub my finger along the top of a table, I feel the table, but I also feel my finger, and in a much more immediate, intimate way.

When I focus on the embodied aspect of inhabiting, I open up a range of experiences that are not available to perception in a more narrow sense, and this means opening to new dimensions of space. Charged with energy and feelings, embodied space carries a kind of knowledge different from the knowledge inherent in the perception of objects.

The focus in embodied experience is different from the focus in perception. To focus in perception is to direct the attention toward or at something. In contrast, if I focus 'on' the belly in an embodied way, I do not direct attention 'at' the belly but 'from' the belly. The belly becomes the center of the looking and experiencing, in much the same way that my intention to do something becomes the center of each action I take to fulfill that intention. And belly-centered space projects space into space very differently than, e.g., head-centered space.

Staying with embodied experience is a practice many people find deeply nourishing. However, it carries with it the risk of sinking into the felt qualities of the experience, losing track of space-projection and space-openness. A more balanced approach is to explore the story that gives meaning to the space we inhabit, starting with the founding story, "Here I am." Any such story has a narrative to it, an *account* of 'what's happening'. But the story is more than its telling, and embodiment is a pathway into that 'more'. By staying centered in the embodied sense of presence as we become aware of the story in operation, we see how we tend to turn stories into narratives—turn inhabiting into telling. If we let this process unfold unchecked, we will abandon the felt knowledge of space in favor of identity, from which space is excluded by definition. But if we can stay within the felt experience, space remains available even as the telling takes hold.

Meditative physical movements that focus on the belly or spine seem especially helpful in encouraging this continued openness to embodiment. However, the same openness is potentially available within any sense experience. In each case, it is the experience of embodied presence (which can engage past and future as well as present) that gives more knowledgeable access to space projection.

There is an interesting tension between opening to sense experience and being caught up in thoughts. The process of thinking, at least in the way we usually engage it, seems to activate a reflex that shuts down sensory experience. It is as though we cannot think and sense at the same time. But that sharp division makes sense mainly from the point of view of thinking. From within embodied experience, thinking is just another telling, which can unfold within a more richly dimensioned space.

## C. The World of the Mind

There are countless kinds of mental objects and experiences, and each arises within its own space. How we can connect with them in the most powerful or revealing way is a vast topic, one which I cannot claim to have explored in depth. Still, a few observations and reflections may be helpful.

A shift in focus from identity to the space that accommodates identity has a strikingly different impact on different kinds of mental entities. When I turn from thoughts to the space within which thoughts emerge, the thoughts simply disappear. When I turn from emotions or feelings to space, the emotions do not disappear; instead, they are transformed, the energy within them released. And when I open to the space within a visualization or a heartfelt feeling such as love or devotion, the visualization or feeling becomes more powerful, more present, more real.

Why would this hierarchy operate? Perhaps because a thought (whether verbal or visual) depends completely on assigned identities; when these dissolve, the thought simply vanishes. In contrast, a feeling, especially one that engages the heart of experience, gains its power—its claim to be real—from a deeper and more encompassing dimension of space. One might even speculate that the more 'real' or fundamental the experience one engages, the more fully it opens into space.

Since the relative unreality of thoughts sets them in opposition to space, other kinds of mental activity give more ready access to space-projection. Perception and embodiment can be the starting point. For instance, I can enter the world of the mind by simply expanding whatever sense of space I am operating with at the moment. As I sit here typing, I am aware of my fingers striking the keyboard and of words forming on the screen in front of me. I can expand that perceptual focus to encompass other objects in the room, the room itself, the room in the building, and so on. As I do so, various associations come up automatically; for instance, the set of books that I notice on my left as I type are not just any books, but books that I helped produce over the course of several years, and a host associations and memories shadows their presence in my field of vision.

The idea here is to open and open and open without ever losing the original focus. Although I am back in the realm of perception, I am engaging a wider perspective in which perception is a kind of mental activity. A powerful example of this kind of expansion into the realm of the mental can be found in the Giant Body exercises in TSK. Exercise 5 asks the reader to regard the outlines of the giant body as "unobstructive" to the reader's knowing presence; next, it suggests adding to the outlines "an ecstatic quality." This instruction suggests that energy and feeling, both grounded in space, are translatable across the gulf between subject and object (mental and physical) in a way that matter, conceived as the opposite of space, is not.

A very different way of exploring space in the mental realm is to start with whatever feels most 'solid'. It may take some experimenting to be able to identify this quite specific sense of something being solid, because we regard what is solid as real, and what is real cannot ordinarily be further questioned. In my own experience, what is most solid is often connected to the self. Its telltale mark is an emotional overtone: dull insistence, stubbornness, anxiety, or the agitation of a restless, thinking mind. At other times, the solid shows up as the vague but impenetrable 'mind set' that prevents me from directing my attention in line with my intention. Simply focusing on this solidity can be enough to melt it, releasing the openness of space. If it does not melt, it may at least change form, and in the transition become more accessible to inquiry. A new solidity, perhaps harder to comprehend or acknowledge, is likely to form almost at once, but still there has been a relaxing and letting go that authenticates space-projection in action.

## III

Another path to activating space-projection involves simply letting go of what arises, accepting it for what it is without accepting its assigned identity as fixed. Such a letting go will engage not only the objects that appear within a given space, but also the 'space/world' within which those objects arise. Since that world centers on a self, this approach ultimately turns on letting go of the self—perhaps the most challenging practice of all.

Walking in the woods, I come to a beautiful glade. The moment I say, "How beautiful!" I invite a reaction of greed, of a self wanting to possess. Unless I let go of that reaction, I have shut space down; or rather, I have filled it up completely with 'my' concerns, wants, and needs. In one sense the difficulty is that I am focusing on 'things'—and my reactions to them—rather than space. But the problem goes deeper. A world that is mine is a world in which space is an inaccessible mystery. When space centers on 'me' and 'mine', on what I want and do not want, it can no longer accommodate. 'I' am out of breath, out of space, out of room.

Tarthang Tulku reminds us again and again in the TSK books that space is around and within, always allowing, always offering freedom. Yet we do not accept that gift. In fact, we actively reject it. Instead of projecting space, we project ourselves. Setting up a position, we set ourselves at odds with space. No wonder we find a space-centered understanding so challenging, so close to being nonsensical. Reacting as selves, we react against space.

We should not let that reaction stop us. In fact, we should regard it as a pointer to something deeply important. When the self says no; when it regards space as empty or meaningless; when confusion and dismay pervade our world, can we open that reaction up? Can we find the space within?

Can space project space into space?

# Transforming Everyday Experience: The Structure of TSK Exercises

Don Beere

Those of us who practice and teach exercises drawn from the TSK vision have often noted their profound effect. The effects seem to occur even when people "don't do them correctly" or doubt the value of the practice. One writer described doing the exercises as running through a planted field and leaving with seeds lodged everywhere about oneself, seeds that later take root.

Here I want to consider the TSK exercises as a whole. I think this can illuminate important elements of the exercises and clarify why they have the effect they do. My hope is that such an overview will be useful for beginners as well as for more advanced students of Time, Space and Knowledge.

## The Structure of TSK Exercises

The TSK exercises have a typical structure. Almost every exercise asks the reader to bring a specific "experience" to awareness and to attend repeatedly to that awareness in a particular way, remaining open to the arising of new awarenesses. In this section I divide this process into its experiential elements, highlighting the most relevant facets.

(1) Bring a specific "experience" to awareness. This process involves being aware of something that either comes up spontaneously (e.g., external events or spontaneous sensations) or is generated (e.g., the Giant Body or diving into time). Although most involve different practices and specific foci for awareness, they cluster into various groupings.

(1A) Attend to some facet of ongoing experience. Examples include observing the past-present-future tinge of all moments (TSK, Ex. 19, p. 174) or the spatial quality of all presenting experience (DTS, Ex. 9A, p. 278). Awareness is not changed intentionally, but rather noticed as it occurs. Experience still proceeds on its usual course, but in its midst I note some quality. An example in everyday experience is when I notice while listening to a beautiful piece of music the quality of the notes, how the cello and piano blend, the phrasing of certain passages, and my emotional response.

Awareness has many different potential facets that can point toward, into, or away from ongoing experience.

(1B) Generate an experience, either in imagination or for analysis/observation. Examples include the Giant Body exercise (TSK, Ex. 1, p. 21) and Stream of Memories (TSK, Ex. 27, p. 256–57). In the first, the instructions are "Entertain in your mind's eye the image of a giant human body . . . very alive and real." The task requires generating the imaginal experience of a giant human body with certain qualities, and then entertaining that image. In the second, the instructions are "Cultivate an awareness of some object, idea, or quality. After a certain interval of time, remember that prior awareness. . . . Entertain an unending series of memory, memory of memory, memory of . . . .") Everyday experience does not usually require us to remember our remembering repeatedly. To do this, I must generate the specific experience.

(1C) Combine A and B by "working with" ongoing experience. In this cluster of practices, I do not actively create an experience (as in 1B), but rather shift my usual way of experiencing. I can, for example, accelerate time and move into the future facet of the present (TSK, Ex. 22A, p. 177): "Return again to a consideration of the past-present-future structure of all moments, select one facet . . . [and] move your sense of located and embodied identity into that facet." I can transparentize a boundary (TSK, Ex. 4, p. 26): "Resume the inspection of the body's structures and systems . . . and open up or enter all opacities and surfaces which delineate a 'structure' or 'thing'." As a third example, I can "go deeply into" some quality such as 'painness': "Let pain into awareness, allowing its staccato throbbing to expand and occupy the whole of consciousness. . . . Cutting inward through pain to painness, you are looking to what arises in experience as experience." (DTS, Ex 16, p. 306–7)

(2) Attend to that awareness in a particular way. Once I have brought awareness to an experience, the TSK exercises generally ask me to attend to certain qualities or aspects. In one exercise I notice the quality of thoughts: "In the movement of one thought to the next, . . . [o]bserve the gravity of thinking in operation" (LOK, Ex. 8, p. 65). In the creation of the Giant Body, I must remain attentive to and create in imagination experience having a certain quality. The body, for example, should be alive (TSK, Ex. 1–6). Later, when exploring the inner structures of the giant body, they should be interactive and vital (TSK, Ex. 2–4). And still later, when working with boundaries (TSK, Ex. 5), they should have an ecstatic quality. As a third example, DTS, Ex. 9A, p. 278) asks me to notice the spatial quality of presenting experience: "Observe what is present and what is happening around you, noting how objects and events appear within dimensionalized space. Focus your observation on characteristics of the sort that we usually connect to space."

To attend to awareness in this way, I must be actively aware, while simultaneously remaining attentive to how I am being aware, so that I can attend in the particular way the exercise requires. In other words, at the same time I am focusing my awareness, I am also aware of that awareness and, in particular, a quality of that awareness. In those exercise situations in which I generate a certain kind of experience, I must also remain attentive to how the experience is currently presenting for me (e.g., with regard to previous examples, whether the memory I am remembering is the same; the giant's aliveness; the bodily sub-system's dynamic vitality; or whether pain is presently filling the whole of consciousness).

Again, such attending is atypical. Most everyday experience focuses on things and the subject or person involved with those things—in other words, on certain substantive details—while ignoring their qualitative interrelationships. But TSK will not settle for that. For example, in "Inventing the Past" (LOK, Ex. 13), we are asked to "pay attention to the evidence meant to validate the accuracy of what is remembered" (p. 107). In "Enlightening Knowledge" (DTS, Ex. 25), we are asked to "simply allow knowledge to know" (p. 337). And in "Source of Thoughts" (TSK, Ex. 11) we are told, "It is not necessary to try to answer these questions; simply attend to them while experiencing the passage of thoughts. Give yourself the time and freedom from common views, so that you may resolve these issues according to your own immediate experience." Similar examples could be taken from every exercise. TSK asks us both to attend differently and to attend to components of experience that are usually ignored.

(3) Repeatedly attend in a particular way. In the "Guide to the Exercises" in TSK, Tarthang Tulku recommends a minimum of 45 minutes practice daily

(p. xviii) and later recommends doing each exercise over several weeks if not months, depending on the exercise. For example, the commentary to "A Mountain Retreat" (TSK, Ex. 15) states, "If this exercise is performed concertedly for five hours a day over a period of three weeks, you will find that 'space' is literally a food that you are eating or grazing on." The commentary to "A Marriage of Sound and Breath" states, "If this exercise is practiced in forty-five minute sessions four times a day, it will assist in an appreciation of 'time' that will in turn allow a more incisive 'knowingness' to appear" (TSK, Ex. 24, p. 188).

What purpose does the repetition serve? In part, the answer pertains to the next facet of the structure of the exercises.

(4) Remain open to the arising of new awarenesses. Awareness must open, or be receptive to, new awarenesses, new insights, new understandings, and so on. The passages already quoted make clear that the TSK exercises open awareness in certain directions or to certain qualities; for example, the discovery "that 'space' is literally a food that you are eating or grazing on" or "an appreciation of 'time' that . . . allows the emergence . . . of a more incisive 'knowingness'."

Paradoxically, in addition to maintaining awareness of what is happening or has been generated, the exercises ask for being attuned to what is new—that is, to what has not yet been understood or experienced. This 'newness' is hinted at by mentioning certain possibilities. The exercises thus invite a constant balance of active-passive, effort-effortless, and intending-receiving.

This helps explain the necessity for repetition of the exercises. It would seem that everyday experience has been constructed with a certain rigidity, filled with assumptions and unquestioned beliefs: a cognitive rigidity, but also a rigidity in perception and in lived experience. Repetition gives awareness practice in constituting experience in new ways, open to aspects that are available but excluded from awareness as it has functioned previously in a day-to-day way.

## Counteracting the Momentum of Everyday Experience

Because ordinary experience tends to be rigidly and deeply structured, it excludes some of the richness and depth that could be available. One of the most frequent ways in which experience (awareness) is structured is via boundaries that separate inside from outside, 'here' from 'there', subject from object and 'now' from 'before' and 'after'.

Most of the time, in the midst of everyday experience, we lose track of the structures that emerge as part of particular experiences. We are preoccupied with our personal issues and goals, and those take precedence over larger and nested contexts. Experiences, whether from 'inside' or 'outside', generate associations and reactions relative to our identity and what is emotionally alive for us at the moment. We tend to become caught 'inside', like a ball in a pinball machine, bouncing from bumper to bumper—from wish, to disappointment, to need, to goal, to emotional response, to thoughts, to reactions . . . on and on.

Bearing this in mind, let us consider in terms of the analysis in the previous section why the TSK exercises have such a powerful effect. How are they able to counteract the momentum of ordinary experience?

(1) Bring a specific experience to awareness. In everyday experience, there is a tendency to slip past experience *per se*, to miss the experience as experience and get caught up in the ongoing momentum of living. The very act of attending to a particular experience as it presents itself changes this. It marks a shift in the usual way we experience.

(1A) Attend to some facet of ongoing experience. Any experience is multi-faceted. We could attend to its edges. We could notice how definite are the borders separating it from other experiences. We could notice the way it begins, proceeds, and then ends. We could notice its density. The range of possible qualities is infinite. Since we tend to miss the experience entirely, we have no chance to investigate these qualities; instead, they can be included in the experience only implicitly. By focusing on a particular facet of the experience, we change all this. We disrupt the automatic way in which we react to experience and and contain it within a particular kind of structure. This lessens the tendency to rigidify or 'freeze' experience.

(1B) Generate an experience, either in imagination or for analysis/observation. In everyday experience we usually generate experience in response to needs or emotions. We often find their presence arising as fantasics or day dreams or brief constructions that seem to present themselves in response to our considering the past or the future. These generated experiences tend to be taken as given. In turn they become another experience to which we can have additional reactions, whether thoughts or emotions.

By generating the experience intentionally, we are initially brought face to face with how we do this—a process we usually ignore. Next, while our everyday acts of generating experience tend to be unfocused, reactions that occur willy nilly in the midst of what is going on, the TSK exercises require us to generate experience in a very focused way, very specifically and also with considerable complexity. As well, since we are not just taking the experience as created but noticing various different facets of it, we tend not to be swept up into the experience, which is what usually happens with experiences in everyday life.

There is another knotty dimension to generating an experience in TSK terms that seems similar to the everyday experience of trying to remember something. In trying to remember, it seems experientially that we open a space and invite the memory to appear, and sometimes it does. Likewise, the emergence of the Giant Body in awareness (TSK Exercise 1) seems to arise out of awareness itself, like a cloud forming in the clear sky. One cannot find a source from which it originates. We invite the giant to come to mind, and slowly (or all at once) the giant does. In other words, generating experience is not something that the subject does, rather it is something the subject allows or makes room for. This very process confronts the subject with its constrictive and limiting role. The subject must act differently to succeed at the kinds of tasks the TSK exercises set.

(1C) Combine A and B by 'working with' ongoing experience. This particular aspect of the TSK exercises is very unusual in terms of everyday experience. Not only are we attending to different facets of experience, we now move into and through those facets. If we are aware of a boundary, for example, we also explore the substance of that boundary. Or, using a different example, we remain aware of the structure of a moment and, with each successive moment's unfolding, enter into the future facet of that moment. The sharp contrast with ordinary approaches to experience seems to carry its own power.

(2) Attend to that awareness in a particular way. As pointed out earlier, usually one does not notice or attend to awarenesses themselves as they present. Most of the time we are swept up and into the flow of experience, the content, without attending to how that content presents itself. Attending to awareness "in a particular way" seems to be an especially effective approach to shifting this pattern. Awareness has many different facets and aspects that tend to be ignored, and the TSK process of attending in a particular way brings some of the variety of experience into the realm of attention. Furthermore, by being able to attend to different aspects of awareness, we discover greater flexibility in the ways we can experience.

(3) Repeatedly attend in a particular way. In the everyday context we almost never repeat an act of attending, since the ongoing stream of experience sweeps us along. To bring attention back to the same focus cuts this momentum. It repeatedly interrupts the automatic tendency to become enmeshed in ongoing experience. Furthermore, by repeatedly attending to the same facet of experience, we are beginning to integrate and to accept as natural an expanded way of being attentive or aware.

(4) Remain open to the arising of particular new awarenesses. In addition to the previous requirements, the individual needs to remain open to the arising of new knowledge, insight or experience. This adds yet an additional aspect to ongoing experience. Despite its usual attentional and intentional structure, attention remains alert for the arising of novelty, that which is outside the previous ways in which experience has been constituted. This way of shifting or, more precisely, opening attention loosens the structures that contain or limit the arising of new knowledge. In relation to this particular aspect of the exercises, we see more clearly the mixture of apparent opposites: active and passive, close attention alongside open receptivity, effort and effortlessness, control and spontaneity.

From a broader point of view, when the TSK exercises are re-viewed this way, their inherent complexity becomes apparent. Likewise, the repeated, different and co-occurring activities of awareness expand and teach it to function afresh and to establish the possibility for new ways of experiencing.

## Transforming Everyday Experience

From an everyday or level-one point of view, some interesting evidence from the psychology of perception illuminates why these exercises work so well. The first pertains to learning to see color or to hear musical notes. Music students who have the capacity to hear tones can develop or sharpen their hearing by juxtaposing two different tones, played for example by two different piano keys. They can learn to hear similarity and difference. Once this is learned, they can learn to play an equivalent tone on a second piano, thereby refining their ability to 'know' what a note is.

A second, more far-reaching example comes from the work of Ivo Kohler, an early Gestalt psychologist who had his subjects wear inverting lenses, so that everything appeared to them to be upside down. There was a time period when his subjects would stumble around and reach in the wrong direction. But after a few days of wearing the inverting goggles, the world as a whole righted itself. The critical variable was action in the world. In other words, subjects needed to move around, attempt to pick up objects, and, in general, use their bodies. The repeated juxtaposition of overtly discrepant cues (visual, motoric, and kinesthetic) eventually lead to the spontaneous reorganization of perception.

Three summary observations from this experiment are relevant. Discrepant cues can spontaneously lead to new ways of experiencing (perceiving). Repetition is necessary for this change to occur. Motor activity or bodily sensing enhances the process.

The TSK exercises invite a similar process to transform everyday experience. Paradoxically, the exercises accept as given the everyday structure of experience and then simply ask the practitioner to observe it—an unusual act. As a result, the constraint or rigidity of everyday experience is gently softened. For example, a frequent relationship observed during the exercises is the pervasive role of the subject in relation to the object of experience. Observation raises a fundamental question: Who is observing the subject-object relationship? It cannot be the subject. Awareness, observing, and knowing cannot be the subject's alone.Then what is going on?

I use this example because it seems so fundamental. But in addition to the subject-object polarity, there are other pervasive distinctions that characterize everyday experience: the boundaries of things, 'here-there', and 'before-now-then'. Precisely these demarcations are the focus of most TSK exercises. Observing them in operation begins to shift their role. Simply attending to what is limiting immediately begins to open those limits.

# EXPLORATIONS

This section of the present volume contains only three selections. Each in its own way is deeply evocative, suggesting something of the riches of the TSK vision.

The first article, by Carl Wittnebert, is a lovely meditation on the nature of TSK practice, both in a group and individually. Several articles that have appeared in earlier sections of the book have asked just what it is about the TSK vision that makes its practice and study so powerful. This article continues that theme. In some ways it complements Don Beere's article, which concludes the previous section, for it draws on a wide range of exercises to explore their effectiveness and speculate on different ways that they could be practiced. But it also functions to introduce the articles that follow, for it suggests how it is that TSK can be practiced collectively, and what this says about the TSK vision itself.

The next article returns to the transcript of a session held by the Bello retreat group that we already sampled in the previous section. Here the focus is not on a specific exercise, or on any exercise at all. Instead, the group engages in a free-floating discussion. The discussion explores the TSK vision, and at the same time it puts that vision into play. The group itself discusses how this is so, though not in so many words. Perhaps in the end these two alternatives—exploring the vision and activating the vision—are not so different.

Another issue that comes up in the transcript has to do with finding the right words for TSK. Several articles in this book discuss the dangers of 'TSK-speak', in which the vocabulary introduced (but also constantly revised) in the books is used a substitute for experience rather than a gateway into experience. The Bello group acknowledge that despite the dangers, words that grow out of shared experience can be the foundation for deeper understanding. The TSK books themselves seem to aim at something similar, though without being able to rely on the kind of shared context for experience that evolves in the course of a retreat. Perhaps it is the reading itself that creates such a shared context.

The final article in this section is another transcript, this time based on an informal discussion between Tarthang Tulku and Piet Hut. Here the issue of vocabulary comes up with respect to communication across disciplines. Is it possible for knowledge to grow in the interchange between disciplines? Is this perhaps an especially fertile field for growth? What factors must be present for such growth to take place?

# Practice, the Practical, and the Particulars

Carl Wittnebert

It was an uncomfortably warm day in September of 1997 when the TSK Association's ten-month program held its first monthly meeting. The San Francisco setting could hardly have been more grand: a two-story artist's studio in a well-renovated Victorian, with fine paintings, some of them inspired by TSK, adorning the walls. The images bobbed and weaved, whispering of the contradictions we embody. Those paintings would provide grazing for my eyes late in the afternoon, when street noise wafted through the open windows. From the roof, there was an intimate view of the skyline. It was fine space, connoisseur's space, and we had hardly begun.

After lunch, someone wanted to take a stroll through the streets, just to get outdoors. But the hostess sank the idea. "I wouldn't recommend that," she said. "Odd things happen south of Market Street, even in the early afternoon. . . ."

So a structure of good space/bad space was established. It would endure. We got very good at having meetings, at the space of meetings. The final retreat was a good example; held high in the mountains, it is etched into my mind as an ode to joy—an implausible mixture of uncontrollable laughter and peaceful solemnity, of abiding in knowing, perhaps even abiding in the vast realm of not-knowing. One could dwell forever in that ambiguity of lightfulness.

Even newcomers noticed, at other meetings during those ten months, that something unusual was going on. "You are all wonderful people," someone said at the end of a meeting. That put it in personal terms. The alternative explanation was that space was opening up.

(By the way, I later concluded that people with an interest in TSK are not wonderful, after all. I had harbored my own elitist views in that regard, but they outlived their usefulness. TSK folks—at least the ones that I know—get depressed, harbor resentment, procrastinate, and are occasionally rude, just like everyone else. It is more accurate to say that TSK almost invariably brings out the best in people. It is still more precise to assert that TSK, by some mechanism not immediately apparent, grants permission for human being to expand to its natural range.)

Whatever the truth about our meetings, the real world didn't notice. The weekends happened, and on

Monday morning, I would find my usual patterns in place. The TSK vision, of course, is not unaware of our lethargy (SDTS, p. 55):

> Knowledge reads the 'can' of time and space and gives it final form. But the order that first-level knowledge discloses reveals more of 'cannot' than of 'can'. Limitations and boundaries seem to hold the upper hand. We cannot extend time, cannot expand space, cannot change patterns or revise habits, cannot open up emotional knots, cannot rearrange the senses. . . . The list goes on. What has become of the power of time and the openness of space in the reality of our lives?

So I thought. Perhaps I was being myopic. Fruitful practice in daily life will naturally differ from practice in an all-day meeting or a retreat. And I began to notice that I was making contributions to others, both inside and outside a TSK context. Not only were the effects of my actions beneficial, but I saw that and appreciated it. My writing, in particular, appeared capable of making a modest contribution to knowledge; along those lines, I sincerely hope that the present article will, in some way, open a door for the reader.

Perhaps something new had begun. Perhaps TSK was beginning to "emerge as a fundamental theme in the ongoing practice of daily life." (KTS, p. xix)

As for my frustration, well, I was in "self-improvement" mode, presupposing an isolated self that—through heroic efforts and the passage of linear time—might yet attain a higher, better way of existing.

As I read TSK, there is nothing wrong with viewing things this way, as long as it does not foreclose other possibilities. This, of course, it has a virtually irresistible tendency to do (KTS, p. 289):

> Taken as a goal, change becomes a possession, something we can 'get' from higher knowledge. But Great Knowledge does not hand out trophies. To hope that it does and hold to our hopes is to hold ourselves back from the fullness of knowing and the whole of our being.

The error in self-improvement schemes, it seems to me, is to base one's notion of path and goal on preconceived ideas, rather than to explore the unnoticed sensations, thoughts, intuitions, and so on, that might show us and our circumstances in a different light. In this vein, might I have profited by inquiring into the frustration itself?

This improvement-orientation, this question of path and goal, is the source of considerable confusion among practitioners. In the beginning, it is quite all right to phrase things in terms of the schemes for improvement that come so naturally to us. Later on, though, things may change (KTS, p. 277):

> True, there may come a point in inquiry . . . when it will seem that major efforts are required to achieve a 'goal', understood in terms of transformation. But these efforts tend to be self-cancelling—without this meaning that they are fruitless. The categories that identify effort as the route toward achieving a specific goal lose

their hold in the face of the recognition that being, as it is, is knowledge.

I take "self-cancelling" to mean that effort and progress come to undermine the very world-view that gives them meaning. It is worthwhile to expend the effort and contemplate the goal, but the significance of both may come to be seen in a different light.

I am writing about practice. There is a tendency to take "TSK practice" as being identical to "the exercises"—the one hundred and eighteen meditation-like attendings, visualizations, confrontations, explorations, perception-bendings, time-compressions, time-expansions, and role reversals that Tarthang Tulku has offered us over two decades. Practitioners sometimes think of these as appropriate only for formal practice, and the books themselves are perhaps not vigorous enough in discouraging that perception, though TSK (p. xviii) states unequivocally that the knowledge exercises are intended to be done in the midst of daily activities as well as in a formal setting. Exercise 30 in the same book, "A Subject-Object Reversal," is expressly recommended for general use, anywhere and anytime.

A similar hint is found in KTS (p. 318):

> As a way of experimenting with [the] capacities of mind, we could allow events and perceptions to arise freely in awareness, in such a way that each object was space and each event time, with the separation between knower and known as knowledge. Patiently cultivated and applied in daily interactions, such a nonstandard way of

experiencing could shift awareness to the quality of experience itself.

"Patient cultivation" does seem important for practice in daily life. The potential for a sudden, breathtaking shift in perspective is more limited than in a formal setting. Those brief intuitive flashes that are a mainstay of practice sessions—like dollops of ice cream that keep us coming back for more, though our deep-seated tendency to grasp at them diminishes their effectiveness—are not so easy to access in the midst of traffic jams, ringing telephones, deadlines, arguments with the boss, and so on. A keen attentiveness to the workaday world may not be without its muted epiphanies, however. Best to go forth with an open mind.

Some who find themselves in crowded or noisy places practice Exercise 24 in TSK: "A Marriage of Sound and Breath," in which an initial attending to the quality of sounds (as opposed to their significance) gradually merges with attention to the breath. More such correspondences can be devised. But anyone embarking on such a course should be prepared to face a pattern of forgetting to practice. This can easily derail the whole enterprise. When we forget, do we blame ourselves? Do we conjure up a "loser" self-image?

If so, some phrases from a discussion of repetitive patterns in KTS (p. 262) apply well. While the act of forgetting seems to be a dead end, could we see it instead as an act of creation, "an active mode of being that renews itself from moment to moment?" Could we view forgetting "at an angle, instead of head-on?" What *is* forgetting, anyway? What is the relationship of forgetting to not-forgetting? Is forgetting the essential background for remembering? Might there bc value in practicing forgetting? Can we even do so?

Here is another approach to daily practice. A traffic jam or crowded store offers opportunities for experiencing the mirror-image relationship that holds between self and world; to see the feedback process whereby a solid world gets built out of nothing, or 'zero'. To see one's mind in a hurry, headed inexorably towards panic, while the world reacts by presenting exactly what is dreaded—yes, this can be instructive. One might even find an eerie beauty in it. The mind sets out agitated and anxious and an obstacle appears; the mind uncannily responds with exactly the wrong thing to do; the world answers, and so on.

It happens very quickly. I have a vivid recollection of making a series of decisions that resulted in my driving through the parking lot of a shopping mall when, in fact, my goal was to be on the freeway, in exactly the opposite direction. The harder and more desperately I tried, the more entangled I became, like one of those dreams where obstacles to progress mushroom uncontrollably. I can't quite account for the power of this seemingly commonplace event, but it seemed to me that something was being revealed to me regarding the nature of mind and world. Perhaps the insight consisted of an unfamiliar intimacy.

A related tack is followed in KTS (p. 337). The text is discussing how a certain particular type of not-knowing manifests as mistakes, frustrations, failures, and misunderstandings:

> Usually mistakes are regarded as unwelcome reminders that the knowledge available to us is limited. Instead of drawing on them as capital available for investment, we see mistakes as signs of insolvency; declaring bankruptcy, we start over, pretending that nothing has happened. The painful consequence is that we make the same mistakes over and over again, letting the limits on our knowledge lead us in closed circles.

The author suggests we can recycle mistakes rather than trash them. We must only "look not-knowing in the face. . . . Like a skilled dancer who incorporates a momentary stumble into the intended movement, perhaps we could apply mistakes immediately in the present, so that the present is directly open to the future." It seems that an improvisational attitude, not tied to rigid expectations and judgments, but fluid and attentive, allows more vitality into the emerging future.

Part Four of *Dynamics of Time and Space* is probably the richest source of material for working with everyday events, especially emotionality. Here is an excerpt from Exercise 12, "Gateway to Presence."

> Reflect on the course that your life has been taking. Without encouraging negative emotions, focus especially on the mistakes and missed opportunities, the broken promises and failed intentions. Let yourself feel the pain that this way of being brings in its wake. Cultivate a sense of caring about this pain, as you might care about a good friend's suffering. Let this feeling penetrate the power of the fabrications you

> ordinarily use to isolate yourself from time's immediacy, so that you can feel time's passing—in the beating of your heart, in the rise and fall of the breath, in the steady flow of thoughts and images, in the hopes and wishes gone unfulfilled.
>
> In practicing this exercise, even immense resignation and despair can be a useful starting point. . . .

I cite this exercise because my resistance to doing it can only be described as monumental. Who wants to think about "immense resignation and despair?" On the other hand, my very reaction may point to some locked-up energy. So how can I approach my stuckness? Might some encouragement help? Why, yes, it could very well make a difference. . . .

This prompts an unusual thought: Could you, the reader, become involved in my practice? From one point of view, our communication is highly structured: I possess snippets of knowledge, which I have reason to think may be of value to you, and I have launched a chain of cause and effect that will result in your having access to my point of view. This seems a bit one-sided, but such is the nature of the publishing business. For communication to flow from you, the reader, to me, the writer, at the time I am writing this, time would have to run backwards, or communication would have to occur outside allowable channels, or both. That is the conclusion one would draw from the usual model.

On reflection, however, we have no way of being certain that this reverse communication is not 'occurring'. When I delete material because it "doesn't sound quite right," thereby making room for other thoughts, am I acting solely on my own impulses? Are your needs and knowledge excluded? Can I even distinguish 'my impulses' from Time, Space, and Knowledge? Within the familiar sequence of writing followed by reading, could there be an unacknowledged intimacy to which you as the reader also have access? Such an inversion of cause and effect still preserves some of the structures of lower-level space, such as isolated individuals. But it definitely has seditious leanings. Does it perhaps bear some of the flavor of the "TSK telling" explored in *Visions of Knowledge* (p. 61):

> We must evoke a different way of using words, a telling that remains a questioning and a questioning that is also an imagining. The reader must join in the telling, like a musician who brings a score alive by playing it instead of simply settling back to listen.

I turn finally to a type of practice with unknown potential that might be characterized as "appreciation of beauty." I have not conducted systematic inquiry here, but have had a few experiences that left me scratching my head. The first step seems to be keeping the mind open in the presence of beauty, allowing for the possibility that something unexpected may happen. I find this most appropriate in an in-between setting: removed from the pressures of daily life, but not constrained to formal practice. The realm of fine art offers one such possibility. As a suggestion along these lines, consider the following excerpt from a newspaper review of "Monet in the 20th Century," a collection of paintings exhibited late in 1998 (Eric Gibson, *Wall Street Journal*, November 17, 1998):

> Since its emergence as an independent subject in the 16th century, landscape had followed the same compositional principle: earth (or water) and sky separated by a horizon line, which organized the picture and oriented the viewer at the same time. Eliminating the horizon was like removing the keystone from an arch. The structure came apart. . . . So in the water-lily paintings, relationships change. The sky isn't overhead but superimposed on the water, and it's viewed beneath eye level, not straight ahead or high up. Above and below, near and far exchange places. Solid matter dissolves into strokes of pigment or a reflection, or hovers somewhere on the indeterminate surface of water. You find yourself looking up and down, at and through simultaneously.

The unexpected outcome here is that a structure becomes more open and fluid—without disappearing.

The TSK books themselves can be approached from this appreciation-of-beauty perspective. One of their stylistic trademarks is an intense lyrical beauty, which easily arrives at artistic expression. Yet the grounding in experience and inquiry steers art toward a very practical end. The results can be spectacular. I once read two particularly haunting passages at a meeting; after I had finished, I began to speculate out loud about their meaning. But the group leader motioned for me to remain quiet; the contrast between my mundane prose and such exalted pronouncements was too harsh. Looking around the room I saw that people were spellbound. This, I think, qualifies as practice.

I end with one of those passages, a bittersweet yet inspiring lyric (SDTS, p. xix):

> Time and space offer beauty beyond measure and power beyond reckoning. When knowledge can free itself from the barriers to understanding that we have imposed, we know this to be so. Within each point of time or space infinite depths are available. In the field of space, the potential for transforming insight beckons, inviting us to claim our rightful inheritance.
>
> Long ago—perhaps when we were children, or perhaps when the human race was still young, we chose a way of knowing that put time and space at a distance. Ever since, we have structured our lives accordingly.
>
> Yet the narrow line we choose to walk, the tunnel into which we compress our being, does not fit our experience or our capacities. Our lives are filled with countless reminders that time and space are more than we imagine. Despite the limiting identities we assign, the radiance of being shines through. The moment we look, a different kind of knowledge is available.

# Practice Day, 9/7/02

## Bello Retreat Group

*In August 2002, several members of the TSK Association got together for a week-long retreat at Bello, California. The same group in various configurations and sizes, had been meeting in workshops, retreats, and study and practice groups for some time. A month after the retreat, the group met again to discuss their experience and continue it. They gathered on the patio of a private home, then moved inside to start a more formal session, transcribed in an edited version here.*

Tom: When we were out there on the patio, there was a quality that seemed like Bello. It was just there all of a sudden! And it was like what has been staying with me since the retreat: the sheer marvel of things. Out there, it was as though the same quality had returned, or like it never left. But when we moved into the house, whatever was there left. How did that happen? It's quite a marvel in itself. First that the quality returned . . . . Now it seems on the verge of coming back.

Jean: Do you think it was from the warmth, or being in nature?

Tom: I think that is in being unknown. The experience is so close and so elusive at the same time. It seemed to touch all of us out there in the patio, and then we came into the known.

Doug: Tom, you started by reading what you had written about the depth of the unknown, and then Lucy, you said something that added depth to it, that made it blossom more. What did you say?

Lucy: Is what we are doing helpful? We're trying to recreate something. That is what we do in out lives: we get an experience that is really there, a direct knowing, and then we try to re-member it, to recreate it, but we never quite get back to that initial experience.

Maybe that sense we are talking about is like a presence. Or you might call it a kind of coherence. Does that seem right?

Tom: Definitely! And what interests me is what presence has to do with what we call the unknown. Does the unknown deepen presence?

Marion: It does if you're aware!

Lucy: The word that comes to my mind is 'abiding' in the unknown—not trying to fix it or change it or run away from it. When we are all 'abiding' in the unknown, there is a relaxed quality. To be able to do that in a group is a really unusual experience.

Tom: It's so easy to dismiss that experience when you are by yourself. But when we are part of it together, it's like a holding.

Doug: What do you mean by "holding?" Do mean the group energy holds that opening?

Tom: My sense is that in our being together the quality doesn't get dismissed.

Lucy: The experience is amplified in the group, in the group presence. It is more tangible, more tactile. It has more substance.

Tom: The external environment becomes part of it. When I'm alone, it's just my experience, but in a group there is also what I would call an 'outside'. And that's somehow related to when we were outside on the patio this morning. It's a different 'outside', but the two are also related. Being on the patio was like being in the redwoods at Bello. When we came in here, it changed.

Marion: Are you suggesting that this kind of open experience replicates what happens in our lives in general? We have a sense of 'closeness'. I wonder if this happens to us all the time.

Tom: I know that things do come all the time, and then I lose them.

Marion: Your saying that makes me realize that is true, but I haven't been aware of it. Just now the past week flashed by, in a panorama, and I realized that different kinds of experiences happened, and I just didn't know it.

Doug: What do you mean when you say they happened and you didn't know it? Did you just become aware *now* that they happened *then*?

Marion: It was the sense of closeness, and then the fragmenting. There was a shift. The place changed, or the people changed—or maybe I changed. It's wild, isn't it? Crazy! Something comes up in me and I am not aware of it, not conscious of it. But it did happen!

Tom: The question comes back to me, "What does that have to do with the unknown?"

Marion: Everything!

Tom: It seems that way to me too. And then, all of a sudden, it seems like everything is familiar again. It is so momentary. I don't usually look at presence as being so momentary.

Lucy: Well, it *is* time, isn't it?

Tom: That's a good way to ask: How is time different in these two different experiences?

Hal: Are you asking, "What was the quality of time out there and the quality of time in here?" A couple of qualities come to me. Coming in here is more of a spatial condensing. The quality of time here is more inward. Outside, the quality of time was more outward, expanding toward the sky and the trees. Coming in here, I felt a kind of drawing inward. There were two different kinds of environments.

Lucy: I think the question of environment is something we cannot appreciate fully. When we are in nature, an aliveness in our being comes forth. We don't think about it but we are affected by it. You could feel thc trccs, cvcn though we weren't thinking about the 'treeness'. The quality of the place was helping us to be 'held'. At the retreat, we were being held by the energy of that place.

Tom: As you are speaking I can feel that quality here. I could feel the hills outside and feel the sky. There seems to be an element of mystery in different environments—the quality of the air, the distant hills. I can feel that here right now as we are speaking. But still there is a difference, being in here. In here, I come back to the known, to "I know this."

Hal: For me it's different. I feel more mystery in here than I did out there.

Lucy: Here it's closed in. Maybe that's something you respond to more. For me, I want to be wild, not closed in. But Tom made an interesting point. He said we can hold that awareness even in here. Once you said that, Tom, I could actually sense the water and the distant hills. I'm connected to that whole environment. That's a neat extension of awareness.

Tom: The mystery element seems to be linked to time. There's a quality that feels more fluid. It might be like touching different times. At least it's different from when there is no sense of mystery. And there may be a sense of denseness. Even the dense quality touches different times. Maybe that's what gives the sense of being familiar.

I'm reminded of practicing the Cloud of Knowing exercise [LOK ex. 16) during the retreat. I became aware of different times being part of my present experience. I hadn't been aware of the interplay of time in all of our moments.

Marion: It is tricky to be aware of this during the day, even as it is affecting you. You are too busy: scrambling eggs, cleaning up, etc.

Lucy: I can imagine you could be informed in some way by time in the very act you are performing in the moment, informed by all of those vibrations or resonances that act through time. You probably couldn't remember every segment, but you could be informed by them. Is time a kind of resonance? Does it have that potential of being a connecting link?

Hal: Just take the quality of change. Out there the light is sparkling on the water in the pool; the wind is moving the trees; I can hear the traffic. There's changing sound, there's changing light, there's changing air that I can feel. In here it's quiet, but I'm speaking not only in terms of 'quiet' but in terms of change. The rate of change in here is much diminished.

Lucy: Isn't change one of the main characteristics of time?

Hal: Yes, changeability. A pervasive ability of time.

Tom: Lucy, you were talking about being informed by time, and you mentioned our being aware of this, but not particulars. I need to get this question straight in my head, since I seem to have more than one thing going on at the same time (laughter). There are two different thoughts. One has to do with how extraordinary life is. I've had a quote from a Buddhist text running through my mind: "Nothing is as it seems, yet everything is exactly as it is!" But beyond that I am seeing that quality of time informing my regular life. There's a way I thought things were and seemed to be, and there's the way we get somewhere . . . .

Hal: I can appreciate that, how wonderful it can be when two perspectives that seemed opposed interpenetrate and play at the same time, and you can appreciate them together, at the same time.

Tom: I think this links to what Lucy was saying about the changeability of time. There seems to be an awareness of time that we include without even trying. It becomes part of our mind stream. There's all this input, coming from all over, not just from where we might have thought. And then there's the linking of these times. So that way much more is possible.

Jean: What do you mean by the linking of time? I appreciate the energy of appreciating time. Is that related?

Tom: I would say that as this awareness of the ongoing informing of time becomes more part of us, there are wondrous linkings, changeabilities, that become more fluid, more pervasive, more present.

Lucy: So there is this simultaneous changeability. For instance, you are outside walking and there is a certain sense of the wind or the temperature on your body. And there are sensations coming from your past and ideas moving toward your future. They are all informing the moment, and if there is an awareness of all of those things, they can link up.

Jean: So this linking can give us more?

Tom: What seems to happen is that we become what is going on. There is nothing excluded. Everything, every moment, every aspect of every moment has this same total informing from different times.

Lucy: Hal, there is something about your energy right now; it's connected to the movement of your head. I'm drawn to that.

Hal: I was thinking when Jean asked about "more," she meant in terms of fullness or completeness, not so much an adding or linking of different aspects. Not quantitative. More complete, fuller, more comprehensive. . . . more meaningful.

Jean: Appreciation lets you get into that.

Lucy: When Hal says "more meaningful," that is connected for me with the way he moves his head and gestures with his hands. So it isn't just the level of knowing, it's in the space between us.

Tom: All these linkings. That's how time is wondrous. We're just touching it, just starting.

Doug: When you say "starting," do you mean that in the sense that we would always be just starting?

Hal: That too.

Doug: It doesn't have to be "either/or."

Tom: Can the beginning and the end be informing at the same time? It's like birth and death: what are the possibilities for human being within that range? So the 'more' is not quantitative.

*long silence: a natural deepening*

Doug: Tom, we were talking about the unknown, and as I heard you describe the glittering wonder of light, how it's coming in, I could see it on your face. You had a glorious image but you couldn't quite connect with it in a way that would let you give us the words. Now, at the Bello retreat, for some reason time got stretched; it got very thin, and then things got very deep—it seemed to happen that way for everyone. For me, I could acknowledge some of those things and put them in words. It was as though something I didn't know had happened. And it validated who I was. On several occasions, it was as though there was an opening in the web, and I could see through there. I had access to something. It was very specific. There were eight or ten of those holes and they are still there. I can see into realms that I couldn't see into before.

What I am wondering about is the unknown. There are things I feel and experience that I cannot verbalize. I don't have language for them. Sometimes in this group we get close to the right language. That is something like what happened a little while ago. Normally there are lots of places we can't go because the space isn't open for it and the agreement isn't there. And I'm wondering how much of what is really wonderful in life doesn't get validated because we can't put words on it. As though if you can't "language it," you don't get it!

Hal: I think there was a quality of sharing that developed at Bello that let us use words that ordinarily might not move us in ways that communicated. When we are close, when we have a sense of intimacy, then the words are more easily shared. It's not only the words, but gestures too. Just the smallest gestures that we share with one another are appreciated and then become amplified in the group. And it's the quality of reflection that goes back and forth between us within the group. When reflection deepens, our communication has many different levels to it—gestures, even feelings. This morning we're possibly approaching, but we're not yet experiencing the depth that developed over those eight days.

Tom: Doug, you were talking about windows or openings that help you describe something going on. That relates to what we said earlier about 'more', and also that phrase I quoted before: "Nothing is as it seems, yet everything is exactly as it is." It might be exactly true that there are things you can't express, but I suspect there is an aspect of time that can link. The quality of the experience right now is not as important. It seems like there is a tremendous depth to those experiences, but possibly that sense of depth could be seen through, so that they are available.

So in the same way, what Hal said about Bello is true: making a link to our experience at Bello doesn't have to go in a sequence. I think there have been times today when we were already 'there'. The link is not necessarily progressive or sequential.

Doug: I would say that some of the things that opened up for in conversation with the group have stayed open. I can notice more things as I go along, if I don't get too 'busy'. There is a lot more available in the world now because as a group we opened the sense of it. We talk a different language.

Tom: It also had to do with how the group 'held', and that is happening today too.

Hal: It isn't an effortful holding. It is more like an embracing. As we all come together we embrace the meaningfulness. The becomes a 'group holding' and holding that is an embrace. It shows a way of sharing, a responsiveness to each other.

Lucy: Doug, you said something about the quality of sensitivity that allowed these openings to happen. Sometimes language has that sensitivity that arises in the environment. It's a little like love. It's a kind of love.

Words can point to something that we all share, but the words themselves can't really describe it. Certain experiences, if you put a word on them, they go "poof!" But there may be a way we can actually share the experience, without words.

Hal: Or with words.

Lucy: Doug is talking about "something beyond words." My sense is that words might be part of the experience. The problem comes when we try to make words into the whole thing.

Hal: We try to treat words as if they are separate or as if they have absolute meaning. They don't. Their meaning develops in sharing.

Marion: They do take on the essence of the context.

Lucy: So the meaning of the word could change in a different context, in relation to all the other factors: the sensitivity, the feeling levels, all the unknowns. The unknown edges . . . . We can almost touch them . . . .

Sometimes I can almost feel them in my solar plexus, and I know I'm almost there. I know I'm really close to this kind of knowing! It's like when the word is just on the tip of your tongue, and in a way you already know it. There is something that is just about to reveal itself but isn't quite there.

Doug: When Hal was talking I missed his word. It was like a hole in the ground. It's not that I need to go back and get it, because there are a lot of things I don't get. But something dropped.

Tom: The word that comes to me is 'beyondness'. It seems that words always have a meaning beyond what they literally mean. There is something more to them. The beyondness is something we become part of as a group. Even as we use words, we are probably 'hanging out' more in the beyondness of them. We're tuning in to more than what is being said.

Doug: Do you notice when you say something that sometimes you didn't know it until you said it?

Tom: More than sometimes. [much laughter]

Doug: Maybe that's a beyondness.

Hal: That was striking, just now. You said, "You don't what you're thinking until you say it," and then it happened among us. We all responded to that. What you were saying became complete in our laughter—in our animation. It was known, but it became even more known.

Doug: Tom, is that part of the beyondness?

Tom: Sure!

Lucy: Because it touched something that we all sensed, that we all experienced.

Jean: We know it is a beyond—we all sensed that.

Tom: What is that laughter? Can we go and stay there? What was that laughter about for all of us? I think it said a lot.

Marion: Because that is what we experience, and we know it!

Tom: Is it just that? Is there something beyond that? For me, there is something so true about what we were laughing about. The marvel is how we get through our day when we don't know what we're thinking until we say it!

Hal: Wasn't it marvelous how that animated all of us? You're saying that you don't know what you're thinking until you say it, and you say that here in this group. You could say it to the wall, you know.

Doug: Or think it.

Hal: But that would be quite different from what happened here.

Tom: It was the strength of the truth of what you said. All we could do was laugh.

Doug: Tom, when you think about life being marvelous or more inclusive doesn't that change the quality of 'right now'? For myself, I like to reach beyond and get something. Then I put words on it and grab it. I transform an idea into something substantial. That's so nice, I enjoy it so much, that I'm afraid it limits me.

There may be a lot going on that I won't reach for because I can't bridge the gap and make it graspable. You don't seem to have that problem. You can reach without worrying about whether you can grasp.

Tom: Well, it's not like 'reaching' and 'bringing'. I think the wonder for me goes beyond linear. It's just "poof," and it's there! And . . .

Lucy: It's gone. [laughter]

Tom: That's the trouble. It's a mystery. A real mystery. You could say the mystery could be dull and sluggish, but that's if you don't really want to know what's going on there. The laughter that we all were part of was part of the mystery.

I think the strength of the truth of what we are doing is really a mystery that we have been communicating. It's the immediacy! [claps hands] That doesn't have to 'go', 'bring', 'come' or add anything to what's there. It's just [claps hands] like that. It's just like the laughter.

Lucy: It does change and open. It has the quality of opening space and changing the space that we are in. Everything lightened up.

Tom: Doug, when you described what you do, you said you get real joy out of it. I think there is something quite marvelous in the ability to do what you do. There is something you see and do that is rich, and there is a kind of coming and a going to it. Nothing more.

Doug: Okay. And it doesn't have to be exclusive, For instance, a while back I just wanted to grab Lucy. I just let that hang out for a while, and then I realized that I didn't want to grab Lucy and cuddle; I wanted to grab her and dance. Once in a while I have that feeling. So realizing that gave it a good form. I can still have that feeling, and now it's perfectly presentable. I mean dancing is just fine.

So once in a while I'll do that with this reaching out. I'll transform it. But you know, it works so well that I'm afraid that it gets to be a habit. I get compulsive with it. I realize it doesn't have to be exclusive; you can do more than one thing at once. But it worries me. I guess it's the instability [clap]. It's there [clap]. It's gone. Just what you and Lucy said.

Lucy: What is this grabbing? That has a very interesting energy to it. Just what is it?.

Doug: Well, it was a kind of male thing, kind of animal. I just looked at it a bit, and it started informing me. If you could find a good way to have access to that energy . . . With men, I think we have about ten percent of our energy tied up in something like that. It's like some people talk about releasing your demons.

Lucy: Like transforming it into something more playful and not harmful.

Tom: When you spoke about grabbing before, I thought you meant grabbing an idea, or an opening. And it seemed there is something rich in that, but you're worried about it becoming a habit.

Doug: It's a way I study something. For instance, I'll create an antagonistic attitude towards an author and work against that. It keeps me engaged, and . . . ummm . . . I lost it!

Tom: Where are you now . . . with that lostness?

Doug: Oh, I shouldn't have lost that one, because it's right here and it's really simple. It's a way that I approach the realm of time and ideas.

Tom: What about stepping back to when you just lost that. Can you stay there?

Doug: If I stay there, the 'where' is that I'm looking at this and thinking, "I can't take fifteen minutes to explain this thing. And can I put it in three minutes? No!" That's where I got lost. In fifteen minutes, I wouldn't get lost. In three minutes, I'm lost.

Tom: So can you stay in the three minutes?

Doug: No! I won't fit into three minutes. For you to go with me, we have to go step by step: 1-2-3-4-5.

Tom: Stay in the three minutes, where we can't go together. Just stay there!

Doug: Let's see. . . . There's a reason, there's a protocol to this.

Tom: No, no, not being a 'but there'. Just 'there'!

Doug: I don't think that's allowable. [laughter] There are different rules.

Lucy: If it is, it's a very complicated rule. There a lot of rules you probably don't understand. You know, we have 15 minutes.

Doug: I don't think you want to go there. Impossibility. See, it's not just going nowhere or nothing that

that you can get lost. You can get lost right in the middle of 'stuff'.

Lucy: You can get lost without going anywhere. [laughter]

Tom: Okay, we laughed, and that's juxtaposed with your going and grabbing something and bringing it back to work with. Can you hold those side by side? And you also said we don't know what we think before we say it. Can you hold that alongside?

Lucy: Maybe other people have experiences of this grabbing in a different way.

Hal: An important part of the grabbing that I saw, Doug, is that you said you grab something and you take it to 'here' [gestures to heart area]. And you said, "I make it my own." So something in "here" is important to the grabbing. It becomes completed in "here."

Lucy: There's the distinction between 'out there' and 'in here'. Two different places. In your gesture it gives me the feeling that there is 'out there' and 'in here', and they are two different places in space.

Doug: You have to divide them to work with them, to get distinctions and put words on them and share them.

Tom: Can you say something about the 'out there'? What is it that you see, before you bring it in? Is it like looking through a window?

Doug: No! Windows are specific to an area you have veiled .'Out there' is so open. There are realms and areas and levels and layers of 'out-thereness' all over the place. You have to focus. So you go out and get something to bring in, to focus on.

Hal: I'm wondering if it's at this point that the 'dancing' takes place.

Doug: A kind of dancing, when I own what takes place 'in here' , 'out there'.

Hal: Where 'out there' and 'in here' come together?

Doug: Right. If it opens 'in here' to 'out there', I know it's real.

Tom: If you don't bring it in here, does it remain 'out there'?

Doug: Bringing it 'in here' is only one process. There's a whole bunch of processes, and that's one of them.

Hal: Were you and I 'dancing' when I gave you that description of dancing 'out there' and 'in here'?

Doug: I think so.

Lucy: It looked like it to me. It was like a link. There was movement.

Doug: That focus just sort of lightens things up, makes them more alive. When I add that focus to 'dancing', that brings it alive. But what if it goes away?

Jean: What goes away?

Doug: That newly found sensation and description of dancing with words with Hal. If you're willing to include everything, then anything can go on, and everything can go away.

Tom: That may be a linking of time, too. What is it that says that time has to take it away?

Doug: Experience.

Tom: It just seems that this informing that Lucy was talking about can possibly inform just the opposite of what is normally informed. So there could be stability—something does not need to go. Could that be an informing that comes from time?

Doug: I doubt it.

Lucy: Is it possibile that there is a kind of stability in space?

Doug: Wow!! [much laughter]

Doug: Sometimes when I go to sleep at night I believe I might not be able to get back 'in'. Not a big deal! [laughter]. Stability in space? That's a novel idea! I didn't think they were compatible. I was just going to look at that. It's like looking both ways before you cross the street.

Jean: Lucy, where did your statement come from? It's an interesting relationship, isn't it? Sometimes you feel like you're saying something that doesn't come from you.

Lucy: It came out of something connected to what Doug was saying, and also out of a kind of experience. When I think about 'out there', it's 'in here'. There are times I really think of getting lost in space. It happens when I'm really inside, deep inside, my own physical being, my own body. For me, that opens up more possibilities than anything 'out there'.

Jean: For me, it's 'out there' and 'in here' at the same time.

Lucy: There is a place that they do connect, but it's through the connecting place that I go 'out there'.

Jean: Do you think it's one direction like that, or is it—

Lucy: It doesn't really matter. It doesn't matter how, but there is always that component to it. Doug talked about windows: it's like windows through the eyes that open a different dimension. That's all the words I can actually put on it, because it really doesn't have words. But, it is a way of going into space.

Doug: When you said that, I could feel what you were saying, and it felt scary. Do you think we're all put together the same?

Lucy: [Laughs.] That's an interesting question.

Jean: What do you mean?

Doug: Lucy describes the process when she goes inside. And you're describing it as 'out there' and 'in there' coming together. From what I see in the world, it does not look like people are put together the same. 'Out there' and 'in here'—different places for different people. We talk as though there is a commonality because words require that, but I don't see a whole lot of commonality between people. It looks to me like we're a pretty diverse bunch. Composed differently, using space differently. Sometimes people will be talking about stuff and I don't get it, exactly. And I believe that everybody else does not get it exactly. We have a habit of saying yes, of agreeing, when we have gotten close. But, it looks to me that we're really a whole lot different.

Lucy: It's possible also that there is something we do share. That's why it's important to really look at a word, because the word can open up spaces where we can share certain kinds of knowing. That's a possibility, isn't it?

Doug: I use that as a facilitating belief. I look at the world that way, sometimes, because I want to find more similarities with people than differences, because it's handier and it's friendly. But not because I think there is any truth in it.

Hal: When you said a while ago that we're all unique, was that something that you were sharing with us that we all understood? Didn't we all know what you were saying?

Doug: Yes, you knew what I was saying. But I don't think that you agree, not at all. I think that there's disagreement there, big disagreement.

Hal: What I'm interested in is how that relates to 'dancing'. When we 'dance' there is something we share, right? There is something going back and forth between us. Was there a 'dancing' at the moment that you said "we're all unique"? Was there any 'dancing' there?

Doug: I think so . . . yes, there was. There is the 'us-ness'. We evolve out of 'us'. That was going on . . . because . . . no, you can't talk about it!

Tom: That's an interesting dance. The differentness.

Doug: Tom, do you think that we are all basically different? It doesn't sound like it.

Tom: Well, that's changed in the last thirty seconds. When you were answering Hal's question about the dance going on, I had a hard time relating to what you were saying, even though it makes a certain logical sense. But, I couldn't rest there. This 'differentness', there's a dance right there! One thing in common is that we really believe that we are individuals, that we are separated from each other. That is something we share, that we dance with. It may not be very intimate, but we are 'dancing' with this shared belief or beliefs or presuppositions.

Doug: So we are all one and we are dancing the 'separation dance'?

Tom: We certainly are all dancing the separation dance. You aren't going to find many people who disagree—who wouldn't dance. When we recognize the dance—

Doug: I can do that! That clarifies a lot. I can actually see, maybe for the first time, how we could all be the same. The quality of what you said, and the way we're playing with it. Before I would think, "Oneness, that's a neat idea; go for that some day." After all, I didn't have any way to hold it. So, Lucy, are we all one, and we're just dancing the separation dance?

Lucy: What allows us to dance, even though we're different, is that we share a rhythm. That's what is interesting about dancing with partners, but we can do it all together too. I'm not dancing a completely different rhythm than you are. That would not work. There's an energy between us that is moving us.

Doug: So at the retreat at Bello's we had a chance to fine-tune our rhythm.

Lucy: Right! I think there's something to that. At the retreat in Santa Cruz, there was the ocean, and we were all 'moving' with the ocean. Then Bello's. The rhythm gets better and better, fine-tuned, so we can become very skillful at moving together, like ice skaters or tango dancers, moving almost like one body, but also moving quite differently. But they are still part of the same rhythm.

Doug: I just love that, when you put it together like that! If we have to have the same rhythm, maybe we are different but doing the 'sameness dance', or we could be all the same, doing the 'difference dance'. Or we could be doing the same at once.

Lucy: And we're fine-tuning it.

Doug: And I can grab that and hold it to me without freezing anything in place. I like it, and it works to orient me, and then I can look into the vastness. It seems I have to be oriented a bit before I can expand. Tom, you seem to take things on faith. You just do it. That seems scary to me.

Tom: What Lucy said about stability in space was powerful. I think there's something to that.

Lucy: I get a sense of it here.

Tom: You get a sense of it for yourself?

Lucy: There is something about this group. The space of this group has a kind of stability. You could also see it as a kind of trust. So it gives stability to the vastness of space or form.

Tom: It's interesting where that trust comes from. None of us is laying it out there, but it develops. We don't 'do' it, but there is a real trust. It seems to have something to do with what is being evoked in us. Something in that is trustworthy.

Jean: Something bigger than us?

Lucy: Something that connects to the unknown, so we can go deep into the unknown. And over time, a sensitivity develops. We're willing to take risks.

Tom: What is the sensitivity to?

*extended silence*

Hal: It might not be 'to'?! It might be a responsiveness. Is that what you mean? Acutely responsive, or more sensitively responsive.

Lucy: Like awakening more.

Jean: The more sensitively responsive I get, the more I develop a trust in the process. And I feel that the trusting . . . is bigger than me.

Tom: For me, trust resonates with the 'truth', just the way things 'are', without altering them.

Jean: How does the trust enter in? There's no question about it!

Tom: Right!

Jean: It creates the trust or you experience trust.

Doug: So you can bring certainty in there.

Jean: With that knowing.

Tom: So that's a sensitive response to . . . sensitive to one's being, maybe?

Hal: With each other, or each other, but including ourselves?

Jean: It seems to magnify the sensitivity.

Hal: Lucy, what are you working on, Lucy?

Lucy: I'm kind of . . . expanding.

[laughter]

Jean: That's how it works, isn't it? We deepen our appreciation and sensitivity and the connection between us, and the resonance, and that leads us to deeper dimensions.

Tom: You say deeper dimensions. I think of deeper truths. We can start with very simple truths. What really is true when you take away a lot of this surface stuff? We were talking earlier about different 'informing', times that were informing. Are they true? Do they touch a fundamental truth?

Lucy: They are like digging up the roots.

Tom: And is this down here [gestures] anything deeper than this up here [gestures]?

Jean: Sometimes words of truth come flying in, and I wonder where they come from.

Lucy: Maybe instead of 'deeper down' we could say expanded, in all directions. A kind of awareness.

Jean: I like the word 'fuller'.

Lucy: It's a language problem again. [much laughter]

Deeper? Expanding? New language!

Hal: That certainly animated us.

Jean: Well, I have an appreciation for that!

Tom: In the commentary to the Big Sky practice [TSK "Mountain Retreat," Ex. 15] it says that opening up space, as you look at the sky, you come to a point where the breath and depth become the same.

Jean: Yes.

Marion: And it moves right into you!

Doug: Tom, you said, "the way things are." But I can't honestly say that things are a particular way.

Jean: How about just for a moment?

Doug: Not really. I mean, I can say that: "the way things actually are." But it's just words. What's behind the words for you?

Tom: What comes to mind again is the phrase, "Nothing is as it seems . . . ."

Doug: Oh, okay!

Tom: "yet things are exactly as they are."

Doug: Okay. Each of the two would be the bottom, the flip side, of the other.

Tom: But, even that . . . I don't know what I just said says!

Doug: You would almost have to say them together.

Tom: At times when I've been confused or not knowing the direction, it seemed that what was always there was whatever the truth was. Truth seems to be a loaded concept for many people, but all I mean is that in your lostness, all you can really say is what is going on. Once you say that, it has already changed, so you can't rely on that either. The only thing you can rely on is the truth. You might be deceiving yourself, but you have to start somewhere. It seems to be a kind of ground, the only ground that leads anywhere.

Doug: Okay. When you said that about the way things 'really are', I was going to go into a deep philosophical thinking about it, but then I realized, "That's a pretty handy thing. I'd like to have one of those."

Tom: It's like TSK. We can get really serious about it. But Tarthang Tulku undercuts the whole vision when he says in the first book that it's only a "particularly incisive read-out on the way things are." Something to that effect.

Doug: It's not the only way, but it's a way. "The way things are." That really had some juice to it—for a moment or two! [laughter]

Doug: I can take that, and the fact that we're all the same and we're doing the 'difference dance', and I can build a whole universe out of that. That's all I need! Two things.

Jean: It's not like there are four.

*Break for lunch, followed by a few casual comments*

Lucy: Were we able to maintain that kind of awareness during lunch?

Hal: I think we did, especially when we all came together at the table.

Tom: Yes. When we came together, I felt that, and I was looking at the quality and at what changed—what was there for me. What keeps coming back is this unknown. Even though I know each one of you as individuals—that's a part of my field—that wasn't predetermining anything.

Hal: Where I most experience it is in those moments when we fall silent. When we don't have the space filled with talk, there is some sort of a fulfillment in being with each other that goes beyond time.

Tom: There's a social part of me that grapples with the silence from time to time, so I have this uneasiness or unsettled feeling. Yet I don't say anything. At the same time there's a space for that in the experience. If there was chatter coming in I wouldn't have the chance to feel uncomfortable, to see my tendencies.

Hal: Are you holding yourself back when you don't say anything?

Tom: No. I look around and see something I could say, and sometime I will and sometime I won't. There is a kind of maneuvering that's nice to have happen in the silence.

Doug: I had a thought about silence. I thought that it's so nice that we don't have to fill it in. Most places it has to be filled in, which is awkward. Did that occur to others?

Lucy: I find it's *so* delicious. You just settle into it, like a big soft place. But something generally does arise at some point that seems to take us out of it.

Doug: (to Lucy) Could you say "*so*" one more time like that.

Lucy: *So* comfortable!

Tom: I want to come back to the unknown again. Even in silence, what's there is, "I really don't know." I don't know if I should talk; I don't know if I should be quiet. I don't know what's going on. There is just an element there that seems to be part of the connection to whatever is happening, to whatever is there. In this case it's all of us. There is a part of me just looking at that. There is that intimacy that we are involved with.

Lucy: That seems like a question. What is this intimacy we are involved with? Your question took us into a deeper place, a silent place again. But now I've opened my mouth and broken it up.

Jean: Well, it doesn't seem to matter if you do speak or you don't.

Lucy: I did, because I thought it was a really interesting question!

Marion: When I was at the last TSK retreat at the Nyingma Institute, we did a little Kum Nye that was profoundly moving. I never felt more intimate with a crowd than I did then. The exercise was just moving your head slowly. I don't know how long it took to go from one side to the other, one movement: 20 minutes, 30 minutes? I was feeling what everyone else felt and it was fine.

Lucy: That must have been the awareness accompanying the movement.

Marion: It wasn't planned or rigid, or programmed. Just the opposite.

*ten minute silence*

Doug: You're talking about the unknown. It feels as if the energy level in the room has been rising up and up and up and up, and now it's so full of energy. It seems like the last five minutes or so it's been rising.

Lucy: Do you feel that Tom?

Tom: I feel that more and more.

Lucy: Now that he's saying that . . .

Tom: Pretty suggestive! (laugh)

Hal: I felt it before you said it, Doug.

Tom: I was grappling with trying to know something . . . . It wasn't working.

Lucy: It's interesting how much time gets caught up in grappling with trying to know something!

*long silence*

# What Matters?

## A Conversation about Science

Tarthang Tulku and Piet Hut

*The following conversation took place at Odiyan on August 11, 1997.*

Rinpoche: Let me start with a question that I have wondered about. What does science understand by the concept "primordial?"

Piet: Well, it depends on the area in science. For example, for astronomy or physics the Big Bang is what is most primordial.

Rinpoche: That is exactly where I have a problem. Objectively, how can you say that time starts, which is what the Big Bang theory seems to say? And subjectively, how can you say, "I know this to be so?"

Piet: I agree. It's a problem. But the Big Bang is only one way that science has of looking at these things. If you look at quantum theory, it says that things begin anew in every moment.

Rinpoche: Either way, there is a problem, because you have to ask: How does the point arise?

Piet: In science that is unknown: there is no answer to that question. We have the mathematics, but not the description. My own view is that because so much is unknown at this fundamental level, if science develops differently, there could be major shifts in what is considered valid scientific knowledge.

Rinpoche: I find it surprising that science would admit so readily to operating on the basis of an unknown. In that case, your starting point is arbitrary. If there is no logic to back up what you say; if you cannot demonstrate that your whole system is reasonable, how can you even take a single step forward?

Piet: Science sees it differently. In science you start from the phenomena, and then you look for explanations. What that means is that you are always taking the process one step further back, and the foundation for what you are doing always remains unknown.

Rinpoche: Let's assume for now that that's valid. The methodology you describe raises an interesting question: Do we have an accurate way of describing appearance?

Piet: That's a different kind of foundation. I would say the mainstream scientific view is that you can indeed arrive at an accurate description of appearance, and that the way to do it is to take everyday appearance and filter out all sorts of irrelevancies, until you arrive at what is really so. But my own view is different. I believe that everyday appearance, complete with our subjective interpretations, is the foundation, and that you have to start from there.

Rinpoche: But wouldn't you agree that everyday experience is built out of concepts? And if so, can concepts really be our starting point?

Piet: Well, science is built on concepts as well. Many people writing about science like to use the metaphor of a house that has been built up in layers. At the bottom is physics, which is the foundation, and on top of that chemistry, then biology, then psychology, etc. At each level, the structures grow more complex. But there is something in this model there is very strange. If physics is the foundation, it turns out that you can change the foundation completely (as happened earlier in this century), and the rest of the structure goes on as if nothing had happened. It is completely unaffected! That is one reason I would say that the real foundation for science is experience.

As to the question whether science, with the concepts it currently applies, sees appearance accurately, that is something we cannot know. Science sees the shadows that reality casts. But reality casts many shadows. As for reality, science does not see it at all.

Rinpoche: Given your perspective, would you say that science can claim to be logical or rational?

Piet: Only internally.

Rinpoche: And likewise, from your perspective, science cannot know reality?

Piet: That's right.

Rinpoche: But doesn't science maintain that there is something real?

Piet: Yes, certainly.

Rinpoche: And doesn't it say that this reality has to be either objective or subjective?

Piet: Well, this is an interesting point. The answer to your question is yes, but exactly on this point science is being forced to change its view. In quantum physics, this comes up based on the relativity of the observer; that is, the fact that what is real is relative to its perception by an observer, and cannot be determined apart from that. In biology there is the interplay between the structures of the brain, in terms of neurons and so on, and experience. In computer science, there is the issue of building robots and other forms of artificial intelligence.

For instance, there has been a revolution in the building of robots in the past few years, based on the recognition that in order to build a robot that can function effectively in the world (say, a robot that could vacuum the floors in an office building overnight), you have to build a world for that robot. So the interplay between subject and object has now been recognized to be a very complicated thing. My own view is that this is a major shift in science. We have had 300 years of science based on objectivity and repeatability. Now I think we are moving into a new cycle—perhaps it will be for the next 300 years—in which science looks to subjectivity and gives up any idea of determinism.

Rinpoche: Your examples are interesting. Can you point to one place in particular where these new ideas are coming in?

Piet: Well, I personally am drawn to the most radical interpretation of quantum physics, which is that there are no things, but only actions and properties. I think there may be some fruitful developments that come out of that.

Rinpoche: Still, even if you focus on processes rather than things, you are committing yourself to certain kinds of understanding. This may be necessary to communicate. For instance, everything goes from the beginning to the middle to the end. If we dropped that structure, how could we communicate? If the middle became the end, there would be no way to tell a story that was complete or convincing.

Piet: But people do not learn that way. A student who is starting to study physics reads various basic principles in a textbook, and in that sense he is starting at the beginning. But in fact he does not understand the meaning or significance of what he is reading. Only later, after he has acquired certain kinds of experience, does he have the possibility to come back and understand what he had read at the beginning. I raise this point because it is a real problem. In our educational system we present knowledge—for instance, the knowledge of physics—as though it were the result of a historical process that went from beginning to middle to end. But that makes knowledge the result of a process of inquiry, which is a very limited understanding of what knowledge really is. The way knowledge arises may follow a very different temporal order.

Rinpoche: Are these different kinds of temporality important to the understanding in physics of reality?

Piet: Time is important only in a specific and limited way. In mathematics (for instance, in geometry), there is no time, but only space. In physics, time does operate, but in a very constricted way, so that, for instance, time is reversible. In biology, time is tied to intentionality, which means that a kind of knowledge comes into play. One way to describe these three levels is in terms of the where, the when, and the how. I hope to write a short article on this point with two biologists. This is one place where it may be possible to introduce some ideas relating to TSK.

Rinpoche: I wonder about this issue of introducing TSK into science. Science can only do science, which means that it has its own field of operation. Now, within that field, science can aim to be comprehensive. But there will always be something outside the field. And perhaps it is just in this 'outside' that the real possibilities for new knowledge lie.

Piet: I would like to reverse that. Suppose that the field that concerns us is TSK. How do you bring someone into that field? In my own experience, I have found that you cannot readily explain what TSK is about; you cannot lead them into the field in that way. It works better when you can do TSK with them, but for that they almost have to be in the field already. It can be frustrating, because sometimes people just do not see why TSK is interesting.

Rinpoche: That model may not work in this case. If TSK encompasses everything, can it really be a field, with an inside and an outside? How can there be anything that is the opposite of TSK? How can there be a meaningful 'unit' that makes up TSK? Even if you say that TSK is reality, what constitutes 'unreality'?

Suppose we just decided that the time-space-knowledge we know is a fiction. In that case, where and how would we take place? Are you willing just to throw away the history of time-space-knowledge?

Piet: Perhaps we have to. If that is my only choice, I would rather do that than agreeing to walk on the crutches of my conceptual structures the rest of my life.

Rinpoche: Again, you may be assuming too much. You say we depend on ordinary structures, as though they were crutches holding us up, but I do not think that is our basis. How can it be, if it is a fiction? How can it hold anything up? TSK is not trying to kick away our crutches. TSK plays the role of the one who reads the fiction, the one who sees the truth of what it is we say, but at the same time sees it as a story. If we think we are relying on crutches, we may want to throw the crutches away, but TSK says that the crutches could never bear any weight to begin with.

If you understand that, something really opens up. It is quite wonderful, because the rigid structures are gone. But it also has a negative side, because our usual ways of doing no longer work. What can we make? Where will we find ourselves?

Piet: I can't answer your questions, but what a wonderful image: to be completely open!

Rinpoche: Yes, completely open! Really a new knowledge! It is like escaping from tyranny. Under the old regime you had no way to know. Now, even though nothing has changed, you are not in chains. Who has lost what? How can we say?

Knowledge is really a great magician, to make everything look as though it is really so. It is like walking for a long time in a dream, but in reality you go nowhere at all. So who manufactures that appearance? I am not asking in order to invite some story or explanation: that is not the point at all. There can be so many stories, like designs woven into a fabric. I am interested in something deeper. Who is telling? Who presents the picture? Perhaps time-space-knowledge does this, and in a very convincing way, so that everything fits together, and the whole account seems accurate.

Piet: If I understand correctly, the only way to get past this set of concerns is to drop the reality of what presents itself, along with the notion that things unfold step by step.

Rinpoche: Step by step is actually one way that TSK operates. But you are right that there is another, more direct way, because you cannot exclude yourself as an expression of time-space-knowledge. You are a part of it. When you do practices, you are TSK. When you look for what is not so limited or narrow, that is TSK too.

Still, this distinction is in some way limited. You may make all sorts of classifications, but the new thought in TSK is that there are not really the kinds of differences we tend to think there are. The differences are there: Knowledge makes them, so we can go rely on them. But the other side is that these variations do not matter. For instance, let us suppose that before I arrive in this place, and after I leave, the place is exactly the same. 'Meanwhile' I am here, which means that something has come up, that something is different. But why make the distinction right at that point? Why make my presence the determining factor? Perhaps for some other purpose we could look at the entire evolution of the human race as one 'unit'; perhaps that could be our 'meanwhile'.

This way of looking can be very challenging. When we look in a mirror, we know that we are looking at something that has depth. Is it the same for the three dimensions of our reality: is there another, hidden dimension that we are not yet aware of? Maybe it helps to see the current, three-dimensional operation as the operation of time-space-knowledge. It does not have to be that fixed. Somehow some kind of system is operating: that's all. The big story becomes a small story.

Piet: Perhaps I can offer a small instance of that from science. Science will sometimes say, "This is not possible." But that "impossible" is not the end of the matter. A hundred years ago we thought many things were not possible, and today they are. Sometimes when I talk to people with a background in science and ask them to move beyond their usual constructs, I use a kind of science-fiction story. I say, "Imagine that one day we hold the view that all of time is one time; that the three times of past, present, and future are just the result of shifts in perspective. What would become of all the distinctions we rely on today?" Occasionally that is effective in helping someone get a sense of how bound we are by our own constructions.

I have another image I use sometimes. I say, "Being is a ball, and space, time, and knowledge are the handles we use to pull it along."

Rinpoche: It's a good model. But . . . Why are there handles at all? Also, why is it that when you pull on one handle, the others come along?

Relatively speaking, it makes sense to say that something had a starting point; that something happened. That's our conventional model, and we should be glad we have it, because otherwise it seems that things would not function well. But once we have that model, we are driven to ask about something: "Did it happen, or didn't it?" But this transition or distinction is not really as fixed as we think. Things proceed very much instant to instant.

It seems the real transition is toward becoming knowledge. But that leads to a different set of questions. Now we need to ask: If something takes shape or form, how does that happen? In one way, of course, there is no need to answer that question: Who is there that needs to be convinced that the answer is right? But there is still enough of our concern with what is real that the question has its own weight. We need data we can rely on. Reality needs a backbone. Why? It just seems to be intrinsic to our human nature.

So if I do care about the answer, I find myself looking for the mechanism of arising or transitioning. I ask myself how we can establish reality, because I am concerned that if we cannot establish it all, the alternative is chaos. But perhaps when I realize this, I have an answer after all, because the very idea of chaos also encompasses the idea of complete order.

Now, is that 'answer' really any help? It may be an answer externally, but what about the inner, intrinsic, incisive knowledge? If we were operating at that level, we might not have to wrestle with reality so much. We would be much more flexible, because we would understand the Body of Knowledge. We would not be restricted to being the bias-stander.

Then again, we don't want to be too anxious to bring about this 'higher' state. As soon as something arises, we are bound by the structure of 'from' and 'to'. But that structure is also part of the Body of Knowledge. Distinctions and exclusions have their source, and we depend on that, but at the same time we can be independent. In TSK you can afford to think this way. There is no need to be afraid.

Most people are afraid of going crazy. But suppose that you lost the meaning 'crazy'—

Piet: That would be terrible!

Rinpoche: Well, it might be a different story. But does that matter? In a dream I may dream I am falling, but no matter what happens, when I wake up I realize that I did not get hurt.

In a sense it has to do with extremes. We may think of the one that is smallest and shortest as also the most narrow, but at the same time it is opening into an infinite angle.

Suppose we speculate that space has no shape. Forms will still manifest, because if they do not, we cannot have 'inside', 'outside', 'edge', and so on: all the things we rely on. But it will all be a matter of perception. Our perceptions are like property that we own, and we make use of them. Up to now, we have used them to support what we pronounce to be so, and the reality we experience could be seen as their embodiment. But perceptions could also be seen as the space that has no shape. And mind could be seen the same way.

It seems we cannot get a divorce from time-space-knowledge. But that does not mean that we have to be defensive about the variations that time-space-knowledge conducts into being. The process just unfolds. Whether I understand what is going on or do not understand may seem to make a very big difference, but if we just do not identify with what is going on, then there is a greater flexibility, and we can change from one side to the other without too much effort. We can act, we can dance, we can treat knowledge very differently. Whatever discipline of knowledge we pursue, we don't need to be the victimized one. Even if we don't understand, it makes no difference. Whatever appears becomes a self-dynamic process.

## POSTSCRIPT BY PIET HUT

This dialog was held in an informal atmosphere, in an ideal setting, a library at Odiyan. I did not attempt to be particularly precise or complete; I was more concerned with conveying a sense of the main spirit of scientific research. Now that I see that this discussion was transcribed and will be published, it is tempting to 'sanitize' it in a more scholarly way, but that would take away the atmosphere of spontaneity. Therefore I am happy to see it here, published in the literal form in which our dialog took place.

Just to give an impression of the kind of additions I was contemplating, let me unpack the fourth paragraph in the dialog. As transcribed, I said:

> I agree. It's a problem. But the Big Bang is only one way that science has of looking at these things. If you look at quantum theory, it says that things begin anew in every moment.

A short way to rephrase and extend my answer would be to say:

> I agree. It's a problem. But the Big Bang is only one way that science has of looking at these things. If you look at quantum theory, it says that things begin anew in every moment, that every measurement heralds a conversion from potentiality into actuality.

And if I were to write something like that myself, with a publication in mind; or even if I were to talk to a journalist for an interview which I knew was going to be published, I might have said something longer like:

> I agree. The Big Bang theory tells us what has happened in time since the original explosion, but it does not tell us what happened before the Big Bang, or even whether there was a before. Fortunately, current research in string theory might address these problems, which until recently were outside the realm of scientific investigation.
>
> As for the question of primordiality, if we interpret that as the question of how something appears for the first time as an actual entity, not caused in a deterministic way from specific causes, we have a surprisingly different answer. What we see around us is then all primordial! Quantum theory tells us that things begin anew in every moment, in the sense that in general measurements herald a conversion from potentiality into actuality, in a way that cannot be explained fully by classical determinism.

Finally, the article I referred to in the dialog, about the where, when, and how, was published in 2000 in *Unifying Themes in Complex Systems*, edited by Y. Bar-Yam (Cambridge, MA: Perseus Books), pp. 259–68. The title of the article is 'Complexity and Functionality: A Search for the Where, the When, and the How," and the authors are Piet Hut, Brian Goodwin, and Stuart Kauffman.

# APPLICATIONS

A recurring theme in the exploration of the TSK vision is its power to affect the way we live our lives. The articles in this section focus on that theme explicitly. The emphasis is on the personal impact of TSK practice in dealing with the difficulties and transitions that everyone experiences, from the routine frustrations of daily life to the deeper challenges that shape our self-understanding.

In his contribution, Donald Beere presents the kind of everyday occurrence—a lost wallet—familiar to all of us. In his case, the loss of his wallet came toward the end of a TSK retreat, giving him a chance to see the relevance of the teachings directly. Perhaps most interesting is his observation that he did not set out to 'apply' TSK; instead, it worked in him spontaneously.

Jocelyn Thompson follows up on this one example with a host of others, some quite small, others related to deep-seated patterns in her life. In and of itself, TSK is not 'psychological' in the usual therapeutic sense of the word, yet her article, like all those in this section, reveals that implementing a TSK understanding can have profound psychological consequences.

The next three articles, all quite short, appeared originally in *Gesar* magazine. Ken McKeon, a long-time TSK student, writes of his experience in applying a specific TSK exercise to the self-image—and felt experience—of having multiple sclerosis. If his brief treatment leaves you wanting to know more, the solution is to look in your own experience. Try the approach he presents the next time you have a cold or are running late while stuck in a traffic jam and see what happens.

Sylvia Gretchen takes on a similar theme: TSK and pain. Drawing on a specific treatment of pain in *Dynamics of Time and Space,* she identifies three or four different stages in working with pain, each connected to different TSK exercises.

Ralph McFall's article, "Consciousness of Stream," is a TSK 'slice of life', a brief encounter that presents the world in a new light. Like Jocelyn Ford, he emphasizes the importance of not knowing and the ways in which not knowing metamorphoses into a far richer experience of knowing.

The closing article, by Michael Gray, takes time as its theme. Taking seriously the TSK insight that time is our lives, Michael reflects on how his life has unfolded and on how new knowledge is connected at the deepest level with new appreciation for time. He offers the possibility that we can draw on the past and future in ways that go beyond the self.

# Opening to Freedom: Deconsolidating Patterns in Experience

Don Beere

The past continuing through the present into the future is how we maintain the consistency and definition of our selves and our world. The same kind of consolidation maintains our 'problems'.

Despite the appearance of restriction and difficulty, we need not be bound or bounded by circumstances, by the patterns that seem to originate 'from' our histories. To allow that freedom requires the ability to track the emergence of those patterns and be aware of how time reads out alternatives. Tarthang Tulku writes explicitly about these process in LOK, pp. 270–75:

> As an alternative to reason and explanations directed at the 'facts' disclosed when the

'bystander' takes up its positions, we can look directly at the way in which the 'bystander' knows. For example, instead of looking at what observation observes, we can inquire into observation as it operates.

. . .

'Open' to investigate whatever appears, such inquiry is 'free' from the obligation to 'recreate' what belongs to the 'already established'. It is not a question of learning new techniques or practices, but of giving up unquestioning loyalty to the techniques and practices that we are already practicing in every moment. A new form of seeing comes into operation.

. . .

An all-accommodating inquiry can be activated through analysis that turns from what is known to the 'coming to be' of what is known. Such analyses lets us drop our unthinking commitment to our own existence. Looking at the way in which the structures we take for granted arise, we can see that our positions are not solid—that we are not what we thought. We see the names, associations, and shared agreements that establish 'the real' and 'the acceptable' in operation. Established patterns are resolved into a more fluid 'patterning'. Investigating how things are joined together in consciousness, we abandon the commitment to our 'positions' and come closer to understanding the temporal order within which 'positioning itself' arises.

I offer here a personal example that helped me see how this process works and then relate it to how we maintain patterns that manifest specifically as problems and more generally as the self and the world.

## THE LOST WALLET

While at the Nyingma Institute for a TSK retreat, two days before I was to leave, I had an enjoyable dinner with a friend. I had with me a clipboard with drafts of an article we intended to discuss. I also had my 'purse', which contained my wallet, including several hundred dollars and my identification, the latter absolutely necessary for me to board my flight home. We arrived early and talked throughout most of the evening, leaving after 10 p.m. when much of the dinner traffic had lessened. Our table had been out of the way, on the side, directly next to the window looking out on the street, my seat such that its back was to the window. I had hung the purse on the back of the chair and placed my papers on the floor to my left at the edge of a low wall.

The next morning, Sunday, I looked for my wallet and discovered it was missing. Immediately I knew what I had done. When I left the restaurant, I had been carrying the clipboard. I always have "something" with me (usually my purse) and, at the time, being swept up in conversation and with something in my hand, thought nothing of having left my purse on the back of the chair. I also 'knew' that no one had stolen the purse. The chair was out of the way. Dinner traffic had almost stopped. The serving staff would have found it upon cleaning the table. That being the case, I was also certain they had the purse and my wallet. I only needed to get there to retrieve it and hoped they were open on Sunday. Without my identification, I could not board my flight early Monday morning.

When I have lost my wallet on other occasions, I have become frantic and agitated, beset by fantasies of having to stop my credit cards, getting substitute identification and so on. The consequences in the present situation would have been more difficult still, as I could not take my flight back home. This time, though, the extensive TSK practice we had been doing proved profoundly useful. I did not *try* to do what I did, nor did I consciously apply what we had been practicing. I simply discovered what occurred as it occurred.

I found that I could watch the mind construct my usual worries. All of them hinged on the wallet having been stolen. Yet I was convinced it was safe. I found the mind beginning to question the logic of that conviction. Alongside the questioning I also noticed my firm conviction of the wallet's safety. I was aware of having a choice as to where to put my attention, where to focus the mind. I could focus it on the questioning or I could focus it on the safety. Whenever the mind began to move into the "questioning the safety of the wallet" mode, anxiety and agitation started up. I began to play with this, to have fun with it, noting that for one of the first times I could remember I truly had a choice as to how I could respond.

Watching in this way, I discovered many different facets of how I create worry for myself. For one thing, not being worried or upset is not a socially acceptable response. I was aware of a subtle social pressure to get upset—not that anyone would really want me upset, but rather that this was a clearly communicated expectation of others. I was also aware of the inconsistency of my response vis-a-vis my own history. In effect, I 'should' be upset.

Since there was a kind of irrationality to my conviction that the wallet was completely safe, it was easy enough to question it. As soon as that questioning started, it began to activate distress. Since I was able to observe this process, I could track how it occurred. Ordinarily I would get swept into the momentum of the thought pattern or the distress and have no awareness of what was taking place, but now there was an awareness of options. Change in my habitual response was associated with how I deployed my attention, how I focused my mind. It was not that I was controlling my anxiety or using relaxation techniques or 'going to the center' of the distress. Distress simply did not occur when awareness was open to alternatives.

As I went about finding the name of the restaurant, getting the phone number, calling the restaurant repeatedly, reaching the janitor but getting no information, and finally reaching the staff, the same options continued to present themselves. Mind could unroll in the old or the new pattern. This arose also whenever I talked to anyone about the whole incident.

At points during the morning, I questioned my self-awareness. Was I in fact as unstressed about this as I felt? Was I fooling myself? Was I engaging in extensive denial or suppression? The accuracy of my self-observation got some validation later in the day, when a fellow participant in the retreat, especially astute and sensitive, shared with the group his experience on having talked to me earlier in the day, knowing that my wallet had been lost. He said something like, "He was so irritating. It was as if nothing at all had happened. There wasn't a trace of worry." That stimulated another group member to remark, "When I talked to him, it was like he was having the greatest fun. He was having a great time with it."

## LESSONS

There are lessons in this for everyone. First, anyone can do this—although I would point out that the 'doing' is really a 'not-doing'. Tarthang Tulku has set forth how to do it in all of his books. Second, to 'do' this kind of thing requires being aware of one's inner processes. Third, what is possibly most important in changing one's response is the focusing of attention. There are moments when alternatives present themselves, and in those moments one need not follow old patterns. Making this choice, however, almost immediately activates those old patterns and brings up other aspects of self and world that are intertwined with them. The tendency and invitation to return to habitual patterns will present itself again and again.

I have described a rather practical and limited example, and the link that it has with TSK might not be obvious to everyone. But to me, given my intense involvement with TSK practices at the time, the link is clear. And despite its being so particular, the situation actually opens up to a much broader terrain. It invites reflection on how we engage in maintaining the consistency and restrictiveness of the self and of the world.

# Blind Spots and 'Not Knowing': Time-Space-Knowledge Practices and Everyday Life

Jocelyn Thompson

## THE ENDLESS LOOP OF THE KNOWN

Have you noticed that the known (what we already know) tends to move us in an endless loop? We often repeat the same thoughts and behavior, wondering vaguely why we're doing it for the thousandth time. Doing it does not bring happiness or get us what we want—so why on earth do we do it? Why are we unable to see any new options? Asking this question, we draw a blank.

Blanks are important. In practice, it is often interesting to abide, not where lots seems to be happening, but where everything seems blank. These 'blind spots' could be mental blocks or areas of subconscious resistance. Could they also be doorways into the unknown, where new options begin to open out?

## BLIND SPOTS

In Douglas Adams' *Life, the Universe and Everything,* a spectacular spaceship lands on Lord's cricket ground—and no one sees it. Why not? He explains that anything that seems too huge or threatening becomes "something that we can't see, or don't see, or our brain doesn't let us see . . . . The brain just edits it out, like it's a blind spot."[1]

Unfortunately, it is almost a reflex to blank out experiences that do not fit our comfortable, familiar—if painfully limited—picture of the world. Our conscious mind really does not want to know about the spaceships. Why? What makes the mind veer off these gateways to the unknown?

Healing practices like TSK (and Kum Nye, also introduced to the West by Tarthang Tulku) invite us to notice such blind spots. The practices constantly challenge our view of the world, the things we believe about ourselves and what we accept as real.

Recently I reflected on the many blind spots that have come up for me in practice, and I found myself thinking about how I approach my work at a computer. Every time I thought about working at the computer with heightened awareness and concentration, my mind bounced off the idea like a superball bouncing off the sidewalk. "Nah," it said "wouldn't work—forget it." Interesting.

Sitting with the resistance revealed a layer of thoughts about what might happen, racing around in worried circles. The main worry is that I do not know what would happen. There is a layer of fear.

I sat with the question: What am I afraid of?

*failure* (I'll try it and it won't work, or I won't be able to work out how to do it)

*embarrassment* (I'll behave in a way that is socially unacceptable)

*radical change to my world view* (then everything else would have to change)

*transformation of self*, or even losing sense of self—going crazy

*something totally unexpected* (something beyond my control!)

So now I have set the intention. I do not know what will happen, but maybe the only way to find out more is to suck in the gut and see—or, as Susan Jeffers says, feel the fear and do it anyway![2]

So I sat at the computer and observed. My body was tense, and thoughts revolved around "so much to do and so little time." I cultivated deep relaxation and attended to the sense of time unfolding ("Commanding Time," DTS ex. 13). When anxious thoughts appeared, time contracted and became jerky and painful. I wanted time to speed up so the stress would end; I also wanted time to slow down so the work would get done. This dilemma intensified my discomfort.

When I relaxed, time spread out and became spacious. Suddenly it no longer mattered how much time was available; I began to enjoy the work, allowing and cultivating a sense of enjoyment, delight, and interaction—as if there was all the time in the world. I felt content to settle into the present, into the task. At the end of the session, I felt refreshed and confident.

There is irony here. In all the years of striving and hurrying, I had never looked for alternatives nor challenged the assumption that bending time was some weird and difficult way to slow down the clock. Yet all it takes is a willingness to relax, let go, and enjoy.

## "NOT KNOWING" AS THE KEY TO THE DOOR

If blind spots are doors to the unknown, then "not knowing" is the key that unlocks our resistance. The first challenge is to become aware of the blind spots, and the second is to stay with awareness of resistance and anxiety, while relaxing into "not knowing."

The proof is how powerfully Time-Space-Knowledge practices such as "Not Knowing as Knowing" (TSK, ex. 31) and "Choosing the Unknown" (DTS, ex. 23) operate in everyday life. Here are some examples, drawn from recent experience:

*Waiting for a taxi.* It is Friday, the end of a busy week, and all I want is to go home and relax. Ten minutes goes by, no taxi. I become aware of waiting impatiently, feeling tense and frustrated, my thoughts fixed on an end to the waiting. It strikes me that I have done this kind of waiting many times before. What about some new options? My impatience demands action, but if I walk away, the taxi might arrive. In a flash, the key point lights up my mind: I do not know when it will arrive. As soon as I accept this, the tension dissolves. I sit and enjoy breathing. When the taxi arrives half an hour later, I feel relaxed and full of energy.

*Waiting for the traffic light to change.* I am late for an appointment. Stress levels start to rise. I pay attention to body sensations, thoughts, breath. Is there any action I can take that will make a difference to when I arrive at my destination? Do I have control over my arrival time? Do I even know what time I will arrive? Maybe I will be an hour late. But maybe I will be on time, or even early. Again, as soon as I admit that I do not know the outcome and there is nothing I can do right now, the tension dissolves. I arrive on time, relaxed.

*Traveling by train in an unfamiliar place.* I have arranged to meet a friend at a station neither of us knows. Tension rises: what if I am on the wrong train, miss the station, cannot find Steve? Again, I realize that I just do not know what will happen—so in the meantime, I might as well relax! On arriving at the station, I see that the intersection where we had agreed to meet is actually an overpass. I am still laughing when Steve drives up a few seconds later.

TSK practice encourages awareness of what we are doing, thinking, and feeling. We can observe the workings of 'what we always do'. This often opens the stuckness, making space for other ways of being in the situation. The above examples all have a happy ending (do we actually influence events by cultivating a relaxed attitude?) But what if I am on the wrong train or the taxi never comes? What if the minor disasters we habitually expect actually happen? Well, maybe relaxing into not knowing makes it more likely we will stay alert to what is really happening (rather than our fears of what might happen) and see more clearly our options for action. Maybe not—perhaps it is another case of "Suck it in, feel the fear, and do it anyway."

## BLIND SPOTS AND CO-DEPENDENCY

"Not knowing" is appropriate in emotionally charged situations as well as in small daily ones.

Recently I arrived home feeling strong, clear, and relaxed after a great holiday—and walked straight into a family crisis. I stayed with it for a time, but at one point, it all got to be too much. I strode to the door, reached for the handle, and stopped. There was a flash memory of the many times I'd done this before, taking flight in mind or body to escape the pain. So I went back and sat down again. As my breath came and went in huge, ragged gulps, there was an intense feeling of terror and vulnerability. For the first time in my life—or so it seemed—I confessed that a lot of the time, I don't know what to do. The relief was huge. Centuries of tension fell away.

This incident was transformative. At a cellular level, I experienced the power of abiding in the unknown, allowing life and feeling to move through me. Ironically, this has led to the feeling of safety that I have been looking for all along—a kind of confidence that whatever happens, I can live it.

I have seen that when I run away from pain, I stop both breath and feeling in order to gain some sense of control and safety. When I stay in the situation, I have to breathe deeply, feel intensely, and admit that I am not in control. In a sense, running away is a desperate attempt to stop living, albeit temporarily—an attempt to stop movement, maybe even stop time. To stay, feel, and breathe is choosing to live.

I am beginning to realize just how much I have been operating under the illusion that I know (or should know) just what to do in any given situation, just what other people should or should not do, what each of us must feel or not feel, and so on. Here is the heart of what psychologists call co-dependency: while I operate within this illusion, I keep seeking safety and happiness through trying to control what other people do.

It simply does not work—and it is unbelievably frustrating to get stuck in that endless loop. Like an overprotective parent, the blind spot seemed to protect me from something scary and overwhelming when actually it hid the possibility of change.

## CHOOSING THE UNKNOWN

The habits of expecting, judging and labeling keep us locked in the known, tied to the illusion that we are in conscious control of everything that happens. In any painful situation, there are usually thoughts such as "it should/shouldn't;""it's good/bad;" "it's too big and scary.". A TSK teacher asked once, "Do you really always have to have an opinion?" And Tarthang Tulku advises, in describing the practice of Kum Nye, "When judgment arises in your mind, use it as a signal to go deeper into sensations and feelings."[3]

What happens if we relinquish expectations, admit that we really do not know what may turn out well or badly, and stay with feelings, even though we may not have a name for them? I have suggested some answers, based on specific examples from my experience. But I can give a more telling example.

Early this year, I realized at the gut level that my life was heading down a known path, repeating destructive family patterns that go back many generations. I had the sense of looking with wide open eyes, horrified, into a known future, with absolute certainty that I could not and would not go there. Then I looked at the other path—the unknown one. It was a leap in the dark. All I had was the certainty that the known path was not an option, and the faith that I would find health and happiness in the unknown.

Facing the unknown, I still had expectations. I expected disaster! I seemed to face the risk of losing everything—marriage, home, family, love, approval, security. So I came up with a compromise. I hoped to move into new ways without losing the best of the old, or that the unknown would quickly become known, familiar, and comfortable. None of these expectations were met. The awesome thing about choosing the unknown is that every step of this path demands that I keep choosing the unknown, moment by moment. The compensations are a greater sense of aliveness, confidence, and, strangely enough, calm.

## CONCLUSIONS

"Not knowing" is central to TSK, which encourages us to stay with areas of resistance, abiding instead of avoiding, with a calm, gentle attitude of inquiry. As we relax into "not knowing," judgments can soften, allowing feelings such as fear to emerge. Resistance and anxiety can soften. We can ask honestly: Are our blind spots doors to the unknown?

As we practice "not knowing," we be able to choose to let go of old beliefs and behaviors. Maybe we can learn new ways and see new possibilities. But make no mistake: We could well be challenged to change. We could even be challenged to make this a way of life, learning to choose the unknown with all its fluidity, ambiguity, and unexpected gifts.

## NOTES

1. Douglas Adams, *Life, the Universe and Everything,* pp. 26–27.

2. Susan Jeffers, *Feel the Fear and Do It Anyway,* (London: Arrow), 1991.

3. Tarthang Tulku, *Kum Nye Relaxation, Part One,* (Berkeley: Dharma Publishing), 1978, p. 20.

# "All Manner of Thing Shall be Well"

Ken McKeon

Four years ago I was diagnosed as having multiple sclerosis, a degenerative disease of the central nervous system. Its cause and cure are equally unknown. My neurologist could offer me no effective medical help. He said that he did not know when I would die or be utterly incapacitated. He advised me to live life to the fullest.

Although his prognosis did not mark me out as an exceptional individual (We will all die; most of us will become seriously ill), it left me feeling profoundly hopeless. The physical and emotional changes I underwent had a strong destabilizing effect; however, they also served to point me towards the substance of my being. As I examined my own situation, I began to appreciate more and more the teachings of the Time,

Space, and Knowledge vision, which I had been fortunate enough to receive and practice.

An important concept of TSK, one that I now kept in mind continually, is that the fabric of time and space does not exclude any thing or way of being. The fact of a serious illness did not cut me off from the most basic wellspring of humanity. Any particular stance or posture I might assume was intimately interwoven with time and space—open and dynamic, never 'stuck'. Though I often felt trapped by the circumstances of my illness, the instant I would allow knowledge to be present, the play of time presented and the openness of space accommodated alternatives of being that I had not even guessed. I found within knowledge a sensitive mindfulness that engendered appreciation, even delight, in whatever situation arose.

Even on a conceptual level one finds relief through such teachings. But concepts can seem superficial or even glib to the individual in crisis. What one wants is thoroughness and depth. One wants the effectiveness of truth in one's life. Being "open," being "dynamic" sounds wonderful. But are these qualities available to us in a real, lived-out sense? Each of us must investigate this question for himself.

The laboratory for my investigation was my own embodiment; the tools were the exercises presented in TSK. The only prerequisite was a willingness to be mindful of my situation, regardless of what that situation was. This willingness was necessarily superficial and limited at the onset, but it was the ground for learning and change.

I chose TSK Exercise 7, the tracking of the mind-body-thought-emotion interplay, for an initial tool. The TSK book indicates that this exercise leads to "a more intimate acquaintance with our embodiment." This is an acquaintance necessary to eventually transcend the limitations of that embodiment.

Doing this exercise led me to uncover the following pattern: I would tell myself that I was a hopelessly sick individual. Emotions of depression and anger would automatically arise. A felt bodily reaction would soon follow, one of lethargy, dull pain, and resistance to any purposeful activity. Thought-body-emotion would combine and support each other, forming an image of failure, weakness, and doom. This image would coalesce and freeze into place as a 'reality', as the last word on the subject. The one that noticed this 'truth' would position himself at a lonely remove from this image. And yet within the exercise no position of thought, emotion, or sensation can be truly removed. Everything is involved in the interaction. All locations are given up to it, even that of the one who is watching.

I found that any willingness to be mindful provided an opening for learning and change. When I considered a seemingly frozen and impenetrable situation, openness remained intrinsic to the situation. It allowed and even encouraged further interaction. The "last word" was found to be inaccurate and premature.

At exactly this moment, the seemingly fixed and solid situation broke-up and re-formed. Both mind and body became energized and balanced. A new, more lively configuration had come into being. There was a felt sense of relief and well-being that persisted to nourish the activitics of a working day.

During the practice session, knowledge had gone beyond the scope of the self. A frozen or set physical circumstance had been faced and appreciated, rather than faced and fought. Time, as the presence of the interaction rather than an abstract index of set duration, had played towards health and liveliness. Space had readily accommodated such a play. I wondered at the time, and wonder yet, if there is not an impulse towards meaning and satisfaction deep within the human condition that is brought to the fore by an open awareness, a comprehensive and penetrating knowledge of our realm. The only way I know to answer such a query is to continually investigate, to continually appreciate and challenge limitations and interpretations.

I am struck time and again by the limitations of the self's usual stance of knowledge, and by the power and grace of time to break up any particular instance of such a stance. The experience has been like one of watching houses being built in one moment and demolished in the next, because there was something further to be exhibited, something more wonderful, more open.

That "something further" may even be what we ordinarily regard as sickness. Sickness too may be enjoyed and exercised in a playful manner. Neither sickness nor health as we conventionally understand them will last. Yet we cling to one and reject the other. What I had begun to sense was that whichever of these ordinarily contrasted states presented itself, I could encourage its inherent value to come forth and could direct it to paths of accomplishment.

Such paths will be different for each individual. For the sick individual they may well take the form of ways to health. The teachings of TSK offer many specific exercises for such journeys. TSK is not a hidden or secret teaching. It is available to us now, and it offers no resistance to us or to our circumstances. Within the TSK vision, sickness reminds us to pay deep and gentle attention to our embodiment. It offers us an opportunity to be mindful of change and value. It may lead us to see the truth and promise in the saying, "All manner of thing shall be well."

# Healing Through the TSK Vision

Sylvia Gretchen

> We may have been living in a barren desert, but when we look with knowledge eyes, we find that the desert has blossomed into a garden of delights. (DTS, p. 233)

Pain is perhaps essential to life. Pain communicates such crucial messages as "Take your hand out of the fire." On a more subtle level, pain, be it physical or emotional, communicates the need for greater knowledge or for a shift in our way of being and acting. This essential message, however, is often lost, clouded over by layers of superficial knowledge and conditioning. As a result of not listening clearly, the message grows more insistent: pain becomes suffering, and we fall prey to seemingly unending cycles of pain producing suffering and suffering generating pain.

TSK exercises teach us how to listen to the central message of pain. They guide us past pain and suffering into deeper, more incisive knowledge. These exercises let pain communicate another crucial message: "Take your being out of conventional knowledge!" When we respond directly, pain as we know it stops and healing at the deepest level takes place.

The TSK exercises related to pain could be classified into three kinds or stages. First are pacifying exercises, which utilize creative imagination of light and color to help us listen more closely to pain. Calming the mind and body, they illuminate our conventional patterns of relating to inner sensations. At this point we begin to see more clearly the stories, imaginations, and other factors we associate with pain.

Second, focusing exercises allow us to recognize each unique moment of time in a way that bypasses entangling stories that connect inner sensations with suffering. These exercises deeply relax emotions and sensations and create a spacious arena for the next stage.

The third stage intensifies experience, using the momentum of pain itself to overcome pain. 'Pain' becomes completely positive:

> The intensity of pain is the fullness of our human situation. Pain activates the emotional senses and brings us to our being in time. It invites the operation of the senses in ways that are not bounded to conditioning, but express instead the fully embodied continuity at the heart of the real. (DTS, p 306)

Putting the power of pain in the service of Great Knowledge, we find a sword of wisdom that can cut through the structures of suffering. We reach a fourth stage, that of transforming. Now we can transmute the energy of darkness, negativity, and pain into light:

> Imagine that your life is unfolding in a positive direction, moving from light to lighter. As light radiates through your awareness, let it expand, touching the dark and painful parts of your own life, as well as the dark chaotic confusion of human suffering. Imagine that all mistaken problems and concerns are released, never to return. (DTS, p. 336)

Renewed and reinvigorated, reminded of the unity of all suffering beings, we discover that

> At the heart of knowledge, 'transcending' is transcended. There is no new state to attain and no old state to abandon. There is nothing specific to be done at all, for the light of transformation of knowledge does not depend on doing or going. (DTS, p. 222)

# Consciousness of Stream

Ralph McFall

Even people who delight in reading the TSK books sometimes wonder what the point is. "How does TSK work?" they ask. "How can I make it come alive?"

Perhaps the example that follows—a five minute 'knowledge' experience—can convey some of the flavor of TSK in daily experience, as I experienced it a while back. It happened quite naturally, without any special effort or even any sense that I was 'practicing TSK', on a Sunday walk on the campus of the University of California at Berkeley.

Attracted by the sound of running water, I stop by a small stream. The sound comes from a six-inch drop-off. As I listen, I am amazed at how loud the sound is.

I notice that I 'know' where the sound is coming from. But as I listen more openly, I realize that the sound is not connected to the thought that carries this 'knowledge'. In fact, my linking of sound and sight ('knowing' that what I see and what I hear are 'the same thing') is also open to question.

Now I try listening differently, 'not knowing where the sound is coming from'. The sound grows richer, more dense—indescribable. From that density, a few notes emerge. I look up and notice a culvert six feet upstream, emptying into the stream. I ease back into not knowing, with no particulars. A new voice emerges from the background, and I notice its source: a feeder stream to my right. Again I let this observation go, relaxing into the ground of not knowing. Soon a delicate texture of new sounds comes to the fore, and I see that it is produced by the ripple of water over some stones near where I am standing.

I turn to leave, aware of a shift as deep as it is subtle. In letting myself 'not know' on a conventional level, I now know much more about where the sound is coming from. At the same time, I am not bound by the answers I have discovered. With no stagnant positions, my mind is still fresh, and I see that the sound really 'comes from nowhere'. Consciousness of stream/stream of consciousness: delightful interplay of knowing and not-knowing.

Reflecting on this small experience later, it strikes me that the TSK vision and the exercises presented in the books are not 'for' anything. They have no prescribed purpose or presupposed endpoints; they simply defuse our goal orientation and open space for freshness and the unpredictable.

TSK is broader than our usual concepts of meditation or philosophy. We might say that it stimulates interactions among all facets of our being. Familiarity with concepts from the TSK vision formed the backdrop for what I experienced while at the stream, but the essence of that experience had nothing at all to do with concepts. What happened was just a free play of knowing, inseparable from not-knowing. Unconcerned with categories or judgments, knowledge can feed knowledge 'aimlessly'.

Of course, having 'had' this experience, it is not easy to let go of the sense that I have accomplished something, or simply the wish to recreate such a positive experience (for the feeling-tone the experience left me with was highly positive). But the wish to duplicate the past or to spell out the resulting insights reintroduces anxiety. It contradicts the free-flowing appreciation that was central in the experience itself.

Still, don't I want to develop, to deepen my study of TSK? How can I proceed with no goal in mind?

Perhaps I can look directly to "not knowing how to proceed," comparing it to how I usually proceed. Instead of looking for a fixed set of steps that I can follow to arrive at a guaranteed result, I can give up the predetermined forecast and rely on the freedom and spontaneity of transformation. I can move from "know-how" to no how.

# Between Zero and One: Exploring Time through Relationship

Michael Gray

*The 'flow' picture . . . is useful for an initial consideration of time, since it relates to both common experience and somewhat higher experiences of a transitional nature. But, we can go beyond this model, seeing instead that 'even on the ordinary level, time does not pass'. From this latter vantage point, time is neither linear nor sequential . . . .* (TSK, p. 136)

The theme of time enters strongly into all of Tarthang Tulku's books. Recently it has occurred to me that each book offers its own special doorway into time's aliveness. At the heart is Time, Space, and Knowledge and the TSK books that follow it, which show that the doorway is there. The other books offer their own unique ways of opening this doorway and entering a greater time. *Knowledge of Freedom* has helped me contact the nourishing fullness of time through grounding in the past, while *Mastering Successful Work* offers a road map that makes the power of the future seem more accessible.

## Spatial Analogies for Time

*Sacred Dimensions of Time and Space*, the most recent book in the TSK series, invites us to reconsider our standard spatial analogies for time. Even its title challenges the ordinary assumption that time, in contrast to space, is one-dimensional. Questions follow naturally: How can time be an unequal partner of space? When everything is wondrously apprehended as a trinity of Time, Space and Knowledge, how can one facet, Time, be no more than a blind groping, dead ahead?

The TSK vision certainly suggests it is possible to enter multiple dimensions of time, as we do with space. It conducts a campaign to discredit the concept of linear time, revealing the pain involved in always relating to the fullness of life only as time gone and time not yet here. It suggests that the well-known limitations of one-dimensional linear time cry out for a greater fullness.

Here comes a surprise. *Sacred Dimensions* moves toward multi dimensionality by heading in the opposite direction. It turns toward the "zero point" as a source of complete openness and inconceivable fullness.

I do not mean to suggest that this is a contradiction. "Zero dimensionality" in TSK terms is not at all pejorative. But in my own experience, zero spells confinement. I associate it with being trapped in the present moment, unable either to recapture the past or believe in the future. Time may flow, but I am confined to a single location on the stream.

Even though *Sacred Dimensions* suggests otherwise, I do not feel ready to jump into zero as something liberating. I feel the need to proceed more slowly, like a caterpillar whose whole sense of reality is limited to the bit of leaf directly under its nose. In the spirit of inquiry, then, I would like to take zero at face value—zero as I know it—and start my inquiry into time there.

## From Zero to One

> Usually we understand these relationships solely from the perspective of the subject, which knows with the mind." (SDTS, p. 166).

Blinded by anger or lust or depression, what do we really experience or know outside the zero-point of the emotion itself? Whether we are talking about time or space, we seem to live much of our lives in a zero-point isolation. Even experiencing the one dimension we think of as time would already be an evolution toward a certain freedom. Is there any way to bring this about, to move beyond the zero-dimensionality of the present moment? Can we at least experience the one-dimensional flow of time directly?

In Euclidean geometry, any two points make a line. But how—if awareness is active in only a single subjective point—can we experience the connection between the points? Is there a relationship linking points of time? What would its nature be? We connect to the past and future through memory and expectation, but these are projections. It seems doubtful that they move us beyond the point-prison of being a self.

What would a relation that allowed us to move beyond the isolation of zero be like? When I ask this question, it is what I know about personal relationships that seems pertinent. For someone suffering from a sense of isolation, relationship means friends and community; for someone who feels cut off from others, relationship can come through treating others as we would be treated ourselves. Everything ever said about compassion, balance, love, and joy enters into the possibility of such real relationships.

On the other hand, we all know what it means to be in dysfunctional relationships with others. Blaming, categorizing, lusting, we reduce to frozen images a reality we thereby fail to see and cannot really connect with. Here seems a point of entry. We can relate dysfunctionally to ourselves as well. By freezing our sense of who we have been, we make being alive the exclusive property of our self-of-the-moment. Doing so, we rob ourselves of the ground from which a vital present might be born.

I do not mean to present this thought as simply a theoretical possibility. It grows out of my own experience over a number of years. And so it makes sense to me to share some of that experience, giving substance to what I am saying.

## Connecting to Time

Growing up, I experienced a painful inability to feel connected to my own life and lifetime. Time felt like a prison sentence that I had to serve out in some inhospitable realm. Looking back, it is obvious that what I needed was some kind of healing breakthrough in how I viewed time.

In early 1983, several things happened that added up to a turning point in my life. My father died, my first marriage fell apart, and I started a new job at the University of New Mexico. A few months into this new position, someone gave me a copy of Tarthang Tulku's *Skillful Means,* and over the next seven years—the period I stayed at the university—I worked with the roots of a painful personal unhappiness, as revealed in the mirror of that slim volume. I also started studying *Time, Space, Knowledge,* a far larger and more intimidating book. Later came visits to the Nyingma Institute in Berkeley—immersions in a world that showed me parts of myself I had forgotten.

During this period I began—for the first time in my life—to value time. The most striking aspect of this shift was a recognition that I could not afford to treat work as less important than the time I spent meditating, studying or playing. This simple discovery multiplied my capacity to be present to life.

Around 1990 or 1991, seven years after the circumstances I have described above, another meaningful nexus of events occurred in my life. My mother died, I remarried, we had a son (my first child), and we moved to Berkeley, where I took part in the Human Development Training Program at the Nyingma Institute.

In the course of this long retreat, an unexpected perspective on my personal past arose. Through practices based on Tarthang Tulku's *Knowledge of Freedom*, several decisive shifts in direction emerged from the past, shifts I had never previously identified and was amazed to discover now. (I describe this further in "Gateway to the Past," published in the Winter 1993 issue of *Gesar* magazine). It came as a remarkable insight that my own past had been eventful and even interesting. Soon after, long-suppressed dreams of what might be began to rise into awareness. I became able to look to the future as something fundamentally open to anything I cared about enough to try.

When my family and I returned to Albuquerque in the summer of 1992, the wish to transform my life—which an intensive retreat such as I had just undergone must seed in anyone blessed enough to participate—led to a new adventure. I linked up with a paraplegic friend, Foster, and we started a non-profit organization, called Friends in Time, to help people with neurological diseases. Ten years later, we have been joined by sixty other volunteers and two wonderful staff. I have also written two novels—that "dream" I put away twenty years ago, which rose up from the past with a new voice.

Now it is 2002. Another decade has come and gone, and my life has remained engaged and interesting. Being in no hurry for new painful losses, such as I experienced at the previous two times of transition, I find myself wondering whether a spirit of rejuvenation can be woven from the cloth of gradual, healthy change. Reflecting on this possibility, and on how I might implement it, I find myself reflecting as well on time.

## Change as the Doorway to Time

As part of the work I have been doing with individuals suffering from neurological disorders, I have learned to expect one or more of my new acquaintances to die every few months. Those who suffer from such disorders are often physically vulnerable. They succumb to suffocation, to serious skin breakdown, and to other ailments. I have no idea whether this continuing exposure to mortality is preparing me for the deaths I don't dare think about, but it certainly brings me close to the reality of change and the truth of our existence in time. That reality is the foundation for my thoughts about relationship and time.

When we dismiss our own past life as not worthy of attention, or carry around unexamined disappointments for what we have done and not done, we mirror the dynamics of a dysfunctional relationship with another. Such attitudes bring swift and powerful vengeance, for if we do not honor who we have been, then that part of ourselves has no way to support our current endeavors. Imprisoned within the narrow compass of a fleeting moment, we move through the world like ghosts. Everything we love is just out of reach, outside the boundaries of our minuscule foothold in time.

The breakthrough in my experience of time that I referred to above is a reversal of this pattern. Looking back at my own past with empathy and understanding, I found I could contact the heart. And when the heart opens, it naturally opens forward, into the future. The zero-point gives way to something much richer.

Here, then is a possibility: Maybe we can deepen our relationships within time by connecting with ourselves, opening up two-way communications founded in greater honesty and openness. The techniques we use to work on relationships with others may be relevant to our temporal relationship with ourselves. We can appreciate with gratitude past brave acts, share responsibility for difficulties not yet surmounted, and get in touch with what life was, is, and will be.

## From One to Many

If we move beyond the zero-point, what options open? If we move only in a straight line, either forward or backward, our choices are limited. But the straight line is seldom the only alternative. Even if the inertia of forces from the past has brought us to here and now, what is to prevent us from striking off from the path that projects itself forward and heading toward what might have been and could still be? If we can develop healthier relationships with our own past actions (have they not given us our present power to change?), we can develop similar relations to the future—our own and the future of others. If this happens, if we inhabit the one dimension of time we already know fully, is it possible that time could open further—open all the way?

In *Knowledge of Freedom*, Tarthang Tulku writes that the senses are two-way channels. The sound of the wind sighing through the trees opens toward the listener, as well as toward the tree branches. We can follow the lead of the senses toward such two-way sensing in every part of our lives. Opening and being opened, we can cultivate a relationship with others and also with the world.

Such friendship frees us from the prison of separation and selfishness. In knowing anything deeply, we are also known, in our own deepest being. Applied to time, this means that we discover our own being in knowing what was and what can be. If regret and shame don't close the gateway into the past, and dread and anxiety don't pull the curtain down on the future, we see that time permeates everything, in all directions and dimensions. Time is our life. It is not some external force bearing down on us from outside.

## The Future: A Time I Once Dreamed About

The TSK vision opens these possibilities in time, and Tarthang Tulku's other books let us activate them. *Knowledge of Freedom* allowed me to listen again to old dreams I had put aside. *Mastering Successful Work* gave me a way to bring vision into daily life. Its practical exercises have been a tremendous resource. They reveal a future not shaped by discouragement and compromise. They teach me how to harness awareness (knowledge of what I care about most deeply), concentration (attention to the possibilities inherent in time) and energy (the experience of time itself), so that I can embark on the journey of my life.

Talking about past, present, and future may seem to affirm the limitations of linear time. Even the claim that the future we want flows from a present in which we open to our deepest dreams still seems to limit us to a single dimension, along which a present moment shapes what happens later in life.

Yet the invocation of dream and vision may also suggest a different dimensionality. Our deepest aspirations do not necessarily come out of a one-dimensional past. At those times when we feel a strong conviction that our lives can be meaningful, we move beyond the limitations of linear time. Put it this way: The concept of linear time may be adequate to account for cause and effect, but is that really the part of life that makes it most worth living?

Joseph Campbell spoke of "living our bliss." TSK has taught me the practical meaning of that phrase. It has revealed that one way to escape the clutches of linear time is to allow our future to turn its face toward where we now stand.

Suppose our deepest dreams do come from our past. Employing the usual spatial analogy that treats time as linear, making room for these dreams in the present would mean opening a corridor between present and past. But now suppose that we turn toward the future in a way that reflects these same dreams. What happens to the shape of this corridor? We cannot turn in two directions at once—not in the same moment. Perhaps we find ourselves traveling a path that is not linear at all, not even sequential: a path that is more like a field, vast enough to contain all three aspects of time. Perhaps we move in time along contours we never previously imagined.

For linear time, there are no forks in the road and also no mountain meadows at the end of a winding path. But that is not our reality. If we can learn to feel less confined within a stream of linear time, such openings and alternatives may become more readily available. For now we are living out an unauthorized autobiography, which has passed itself off as the final word on who we are, but perhaps we can look at our own past and future in more open ways.

Practicing TSK, trying to embody it, has helped me to consider this as a real possibility: that we can make room in time, and in our lives, for what matters most deeply. Our dreams, our present reality, and our possibilities: are these the sacred dimensions of time?

Here is a final thought. I may have been too modest in my imagining till now. Suppose our dreams come from elsewhere than our own personal past—say from the sky above us or from the earth under our feet. I say this without knowing quite what I mean, but I do know this: Before linear time could drink this in, it would have to shed its skin.

I do not think we need to understand these deeper dimensions of time to get underway. We can use the knowledge available now. Here is a single example: We can open our heart to someone we used to be. There is effort involved here: For our old enthusiasms and hopes to travel forward to where we now stand, the present moment may have to expand; for subjective awareness to embrace this greater reality, it may have to stretch. But the potential reward is great. When an old dream arrives, bringing new (but familiar) life to the barren isolation of a cut-off present, we gain more than a new memory. Joining forces with a part of ourselves that had to journey through time to reach us, we gain the power to break the links in the chain of lonely moments. Having met this time-traveling emissary from our past, we can step into the open field of our own life. And when we recapture the past in this way, we may discover that it was really here all along, waiting patiently to be invited to the feast.

# CONTRIBUTORS

Helena Antonova began studying the works of Tarthang Tulku in Moscow in the late 1970s, when the books could circulate only in unauthorized *samizdat* manuscripts. She participated in several translations.

Donald Beere is a clinical psychologist in the Cincinnati-Northern Kentucky region and Professor Emeritus in Clinical Psychology at Central Michigan University. His specialty areas are trauma and dissociative disorders, and his orientation is through phenomenological philosophy. He has taught TSK at the university level and chaired a dissertation that is the only empirical research on TSK. He teaches TaiJi and is an artist and unpublished novelist.

Bello Retreat Participants

Doug Kent has been taking part in TSKA workshops and retreats for the past three years. He began studying the TSK books on his own in 1985.

Hal Gurish: see separate entry

Lucy Lewis has been active in TSKA for a number of years. She teaches a class in movement that some members of TSKA participate in weekly.

Jean Loyd has been participating in the first TSKA study group since it formed in 1996. She takes part in TSKA workshops and retreats, and has twice traveled to TSK retreats in Europe led by Hal Gurish.

Tom Morse: see separate entry

Marion Skoss has been taking TSKA programs since 1998, as well as classes at the Nyingma Institute.

Charles T. Davis III is Professor of Philosophy and Religion at Appalachian State University. Although specializing in New Testament Studies, he has also published articles and reviews in the fields of American Religion, Computers and the Humanities, Philosophy, Psychology and Religion, and Buddhist Studies.

Ruth Fellhauer has spent more than thirty years as a member of the Nyingma community. After being active for many years at the Nyingma Institute and Dharma Publishing, she is now a resident of Odiyan.

Meindert Gijzen's long-standing inquiry into core questions about being human led to academic study and training as a clinical psychologist and then as a psychotherapist. At present he is engaged with treatment and management in the complex field of psychiatry, addictions and criminal behavior. He is continuing his personal study and practice of Buddhist teachings, always pursuing the main essence and unity.

Michael Gray lives in the mountains east of Albuquerque, where he helps his wife deliver the US mail and works with people with multiple sclerosis and Lou Gehrig's disease. His two novels remain unpublished.

Sylvia Gretchen teaches TSK workshops and retreats at the Nyingma Institute, where she is co-dean. Sylvia has been a student of Tarthang Tulku since 1969 and has been an editor for Dharma Publishing She has worked on the Yeshe De Text Preservation Project since 1983 and is a member of the Yeshe De translation team.

Hal Gurish participated in the first TSK training programs led by Tarthang Tulku in 1976 and 1977, and began teaching TSK in 1978. He teaches long-term retreats in America and Europe.

Piet Hut is professor of astrophysics at the Institute for Advanced Study, in Princeton. He is involved in the project of building GRAPEs, the world's fastest special-purpose computers, at Tokyo University, and he is also a founding member of the Kira Institute.

Ralph McFall has been a member of the Nyingma community for more than twenty years. He has served as a dean at the Nyingma Institute and co-director of Nyingma Centers. He lives and works at Odiyan.

Ken McKeon is a teacher and writer living in Berkeley. He has studied and practiced TSK since its publication. He currently teaches TSK at the Nyingma Institute.

Tom Morse began studying TSK in 1978. He has been co-teaching a TSK class at the Nyingma Institute and has been active in TSKA since its formation.

Lee Nichol is a writer living in New Mexico. A former student of J. Krishnamurti and David Bohm, he occasionally teaches TSK workshops, classes, and retreats at the Nyingma Institute.

Jack Petranker has been the editor or co-editor for all the books in the TSK and Perspectives on TSK series starting with *Love of Knowledge* (1987). He has begun publishing articles in related fields, and in 2000 founded the Center for Creative Inquiry.

Ronald E. Purser is a Professor of Management at San Francisco State University, past chair of the Organization Development and Change division of the Academy of Management, and co-founder of the International Network for Time in Management and Organizations (INTiMO). He also serves on the board of the Center for Creative Inquiry.

William Soskin was a research psychologist, university professor, and founder of a program to deal with drug abuse among adolescents. He taught TSK seminars in the 1980s and spent time on retreat at Odiyan.

Jocelyn Thompson teaches yoga and pranic healing in Brisbane, Australia. She studies Kum Nye and Time-Space-Knowledge at Turiya School of Yoga, founded by the late Maaida Palmer.

Tarthang Tulku was trained as a lama in the Nyingma school of Tibetan Buddhism, studying with some of the greatest living masters before leaving Tibet in 1958. He is the founder of Dharma Publishing, Odiyan, the Nyingma Institute, the Yeshe De Text Preservation Project, and the Annual Ceremonies for World Peace at Bodh Gaya India. His books have been translated into at least fifteen languages.

Carl Wittnebert is a financial journalist. He began his study of TSK in the late 1970s.

# PUBLICATION DATA

The following articles have been previously published in *Gesar Magazine*. In most cases they were lightly edited or revised for use in the present volume. They are listed in the order of appearance in the volume.

"'Well Now, What Did You Think?'" *Gesar* 4(3) (1977).

"Child's Play." *Gesar* 11(1) (1992).

"Celebrating TSK." *Gesar* 13(2) (1996).

"Light and Good Fortune." *Gesar* 13(1) (1995).

"Prior to Theory or Practice." *Gesar* 14(1) (1999).

"Playing in the Fields of Being." *Gesar* 12(3) (1994).

"Body of Knowledge." *Gesar* 10(3) (1991).

"Focal Setting." *Gesar* 7(1) (1981).

"'All Manner of Things Shall Be Well'." *Gesar* 8(3) (1985).

"Healing through the TSK Vision." 12(4) (1995).

"Consciousness of Stream." *Gesar* 12(3) (1994).

# THE BOOKS OF THE TSK VISION

## TSK Series

**Time, Space, and Knowledge:**
**A New Vision of Reality**

*"An integrated, natural intelligence, unfragmented into reasons, emotions, sensations, and intuitions, is our greatest treasure."*

Inviting readers on a journey into the fundamental principles that govern human experience, *Time, Space, and Knowledge* presents our current understanding of reality and human nature as one particular 'focal setting' on experience. As an alternative to the dichotomies and positionings of conventional thought, it offers a unified vision that unfolds from the interplay of space and time with knowledge. In the TSK vision, boundless creativity emerges as the human birthright, offering complete freedom and ever surprising opportunities. Thirty-five experiential exercises, closely integrated with the text, turn theory into direct experience. This book reveals with remarkable clarity that time and space can manifest differently, with lasting implications for what it means to be alive. 306 pp., 35 exercises, 17 color plates, ISBN 0-913546-08-9. Also available in Dutch, German and Italian.

**Love of Knowledge:**
**Foundation, Inquiry, and Transformation**

"Love of Knowledge *proceeds by returning to a point 'before the beginning'—tracing out links between space, time, and knowledge and our ordinary understanding."* — Preface

*Love of Knowledge* combines experiential exercises, graphics, poetry, and logic to challenge conventional claims to truth and awaken the true "love of knowledge" ("*philo-sophia*") that has always guided visionary insight and healing wisdom. Starting with an inquiry that identifies and analyzes the structures of conventional thought that give rise to "technological" knowledge, the book goes on to question the ways we generate and activate our ordinary understanding. There is a special emphasis on the different identities and roles played by the self and how they are set in place. Encouraging inquiry free from all positioning and conditioning and "committment without commitment," *Love of Knowledge* holds special interest for readers trained in Western approaches to knowledge. 411 pp., diagrams, 50 exercises, glossary, ISBN 089800-138-2. Also available in Dutch.

**Knowledge of Time and Space:**
**Opening the Heart of Creativity**

This work celebrates the power of time and space to free us from all limitations. Light and dancing, deep and provocative, the text communicates the pleasure of unbounded inquiry. Like an ever-fresh work of art, it reveals in over 100 short chapters the facets of an organic vision, sweeping the reader up into an improvisational adventure that explores the flashing aliveness of knowledge. 499 pp., 8 exercises, ISBN 0-89800-206-0. Also available in Portuguese.

**Dynamics of Time and Space: Transcending Limits on Knowledge**

*"In the beginning I thought about time and space and knowledge more or less as everyone else does. Now . . . I have found that I am truly not separate from space and time and knowledge, and it seems to me that this is true for all of us."* — *Preface*

Friendly and supportive, helpful in surprising ways, yet also deeply challenging, *Dynamics of Time and Space* draws readers gently into a new way of understanding time and space, casting experience in a new light. An extended treatment of the relation of time and pain has been of great benefit to numerous individuals. Many readers find *Dynamics* the most accessible of all the TSK volumes, a fresh point of entry into this transformative vision. 339 pp., 25 exercises, ISBN 0-89800-266-4. Also available in German.

**Sacred Dimensions of Time and Space**

Although *Sacred Dimensions*—the most recent work in the TSK series—continues and expands themes from the earlier volumes, it also offers a wholly new approach. Starting from the most fundamental fact of all—our presence in the world—it builds up a new "geometry of meaningful existence," in which zero, ratio, cones, and the number 16 all take on remarkable significance. The presentation as a whole represents a challenge to our sense of how knowledge can emerge, even for readers who have found inspiration in the earlier titles. More than forty graphics, included in an appendix, guide the reader toward visualizing a new way of being. 261pp., ISBN 0-89800-360-1.

## Perspectives on TSK Series

**Dimensions of Thought: Current Explorations in Time, Space, and Knowledge**

Published in 1980, this two-volume set of thirteen essays by early readers of *Time, Space, and Knowledge* includes contributions from a number of well-known philosophers, phenomenologists, and psychologists, as well an extensive interview with Tarthang Tulku that offers one of the best short introductions to the TSK vision. Not surprisingly, these early essays tend to emphasize comparisons with other thinkers and approaches. Volume 2 also contains numerous short reports by readers. Each volume is available separately, Vol. 1, 237 pp.; Vol. 2 , 293 pp. ISBN 0-913546-79-8 (vol. 1); 0-913546-80-1 (Vol. 2).

**Mastery of Mind**

This set of eight articles by long-term students of TSK shows a shift away from theory toward applications. These are essays in the literal sense of the word: attempts to come to terms with fundamental issues. Their integrity and their respect for the TSK vision are reminders of how much the vision has to contribute to knowledge in all its aspects. 380 pp., paperback ISBN 089800-245-1.

**Visions of Knowledge: Liberation of the Modern Mind**

The principal parts of this book were written by Tarthang Tulku, in acknowledgement of the work that went into *Mastery of Mind*. Exploratory, dynamic, and richly suggestive, *Visions of Knowledge* draws on the power of imagination and speculation to transform standard ways of knowing. Light and accessible, it invokes a transforming intelligence that delights in the riches available within ordinary experience. Part One offers sixty questions, with related commentary, as tools for inquiry. Part Two consists of short, related essays. Also included are comments by students and an analytic outline of *Knowledge of Time and Space*. 177 pp., 4 color plates, appendices. ISBN 08900-258-3.

**Light of Knowledge: Essays on the Interplay of Knowledge, Time and Space**

Eleven articles, including an illuminating essay by Tarthang Tulku on "Geographies of Knowledge." Mostly written in an academic style, these articles suggest deepening appreciation for the TSK vision and the role it could play in our culture. 444 pp., ISBN 0-89800-287-7

**A New Way of Being**

Twenty-three practitioners offer an inviting introduction to the TSK vision by showing how it unfolds in their own practice and thinking. Some apply TSK to their professions or fields, others engage in dialogs transcribed from study groups. Each article conveys a commitment to go beyond the limits of conventional knowledge, language, and concepts. 292 pp., ISBN 0–89800–356-3.

**A New Kind of Knowledge**

An engaging collection of seventeen articles by students who approach the underlying relation of time and space to knowledge from a wide range of perspectives, using a variety of styles. The tone is both friendly and invigorating. The articles contain penetrating reflections on knowledge, time, and experience, as well as exploring practical applications in psychology, teaching, business, and time management. 300 pp., ISBN 0–89800–357–1.